Modern Aspects of Electrochemistry

Volume 62

Series Editors

Constantinos G. Vayenas, Department of Chemical Engineering, University of Patras, Patras, Greece

Ralph E. White, Department of Chemical Engineering, University of South Carolina, Columbia, SC, USA

This well-respected series commenced publication in the early 1950s. Over the years it has earned an excellent reputation by offering high quality reviews of current research into all areas of electrochemistry and electrochemical engineering and their applications. The international magazine Chemistry & Industry notes that this series "deserves a place in electrochemistry libraries and should prove useful to electrochemists and related workers." Starting with Number 43 in the series, two features have been introduced:

- Each volume is thematic and focuses on a new development in the field or reviews a well-established area that is regaining interest.
- Each volume is guest-edited. As in the past, all chapters are topical reviews.

Subjects covered in recent volumes include:

1. Modeling and Numerical Simulations
2. Progress in Corrosion Science and Engineering
3. Electrodeposition: Theory and Practice
4. Electrocatalysis: Theory and Experiments
5. Interfacial Phenomena in Fuel Cell Electrocatalysis
6. Diagnostics and Modeling of Polymer Electrolyte Fuel Cells
7. Applications of Electrochemistry and Nanotechnology in Biology and Medicine

Shriram Santhanagopalan

Editor

Computer Aided Engineering of Batteries

 Springer

Editor
Shriram Santhanagopalan ⓘD
National Renewable Energy Laboratory
Golden, CO, USA

This work was supported by the U.S. Department of Energy's Vehicle Technologies Office

ISSN 0076-9924　　　　　　　　ISSN 2197-7941　(electronic)
Modern Aspects of Electrochemistry
ISBN 978-3-031-17606-7　　　　　ISBN 978-3-031-17607-4　(eBook)
https://doi.org/10.1007/978-3-031-17607-4

This Springer imprint is published by the registered company Springer Nature Switzerland AG
The registered company address is: Gewerbestrasse 11, 6330 Cham, Switzerland

Foreword

Mathematical modeling has become a critical component of battery development. New batteries are being developed for a host of applications from consumer electronics to electric vehicles. Consequently, the role of mathematical modeling in designing and building these new battery cells, modules, and packs has become critical. Physics-based battery models are being used extensively to determine design factors that limit the performance of a cell, factors that govern degradation of a given battery design under prescribed operating conditions, and to estimate the remaining useful life of cells and packs. As larger form factors emerge, these models will have to be retooled to incorporate local variations in temperature, resistance build-up, and other heterogeneities. The need for rapid prototyping and scale-up in record turn-around times demanded by the electrification of vehicles has prompted development of a new generation of software tools. These virtual design platforms integrate geometry, performance, life predictions, and safety assessment into a comprehensive suite of mathematical modeling tools that Industry can readily adopt to simulate different case studies to match form factors and chemistries to target applications.

With feedback from various participants in the supply chain, the U.S. Department of Energy initiated a dedicated effort to deploy advanced simulation tools for Computer Aided Engineering for Electric-Drive Vehicle Batteries (CAEBAT). The program was built upon years of phenomenological model development efforts by world-class teams comprised of technical experts on mathematical model development, software companies, and end users with active collaboration from Industry, Academia, and several National Laboratories. Since then, these tools have been extensively licensed around the world. This volume is a compilation of some highlights from this effort.

This volume highlights the rampant growth of the battery modeling industry giving practical examples of the different toolsets available in the market and how these have been utilized progressively by the scientific community. There are instances of current-density distribution measurements that have dominated the literature on classic electrochemical engineering for decades, and how these are relevant to evaluate design features in commercial cells of different formats using detailed multidimensional performance models. With access to higher resolution measurements, quantifying the effects of local heterogeneity at the electrode level is no longer deemed a mathematical curiosity. Advances in both experiments and mathematical models are discussed with detailed examples. There are several instances of deploying mathematical battery models for commercial use covered by experts in Industry. Starting out with the Multi-Scale Multi-Domain models that decouple the geometric design of the cell from chemistry-related parameters, this volume shows practical examples of commercial cell and module design, providing extensive validation across length scales. Calibration of model parameters, especially for difficult to obtain mechanical and thermal abuse cases, is discussed at length. The complex interplay of mechanical, thermal, and electrochemical phenomena in determining a given cell or module design is highlighted with practical examples. Gaps in understanding battery safety are also called out using detailed phenomenological models. This volume concludes with an emphasis on the high-performance computing aspects of battery modeling, which is an emerging trend. This chapter provides tangible examples of scale-bridging and discusses opportunities for data science and machine learning in this data-rich discipline.

Overall, mathematical battery modeling has come a long way in terms of the sophistication in how the technical details are represented, ease of adoption, as well as widespread use. As newer cell formats, chemistries, and applications emerge, this field has grown exponentially to help navigate gaps in our understanding of the material limitations for example. With increased emphasis on virtual development and monitoring platforms, the mathematical battery modeling community has evolved drastically to enable deployment of real-time fault-monitoring systems. Reliability improvements assisted by extensive mathematical modeling and analysis of large datasets using an automated pipeline of software tools are becoming the norm. This volume provides a comprehensive overview of the state-of-the-art in mathematical battery modeling with an emphasis on practically relevant solutions.

Columbia SC, USA   Ralph E. White

Preface

Building a battery has remained as much an art, as it has been precise engineering through several generations of cell formats, applications, and chemistries. As with any intricately well-designed system, the scale and range of complexities at the cross-roads of multiple disciplines provides for a rich set of challenges to tackle.

A carefully thought-out program was initiated by the Vehicle Technologies Office at the U.S. Department of Energy, with the intent of employing simulation tools to support technically rigorous and cost-effective design of automotive batteries. Since their inception, the Computer Aided Engineering for Electric-Drive Vehicle Batteries (CAEBAT) umbrella of projects have significantly influenced battery manufacturers, original equipment manufacturers (OEMs), and academic researchers alike. The suite of tools deployed under this effort have been focused on minimizing the build-and-break cycles associated with battery design while shedding light on the nuances of process engineering and how they relate to battery performance in the long run. This volume attempts to provide an up-close look at several aspects of modeling the performance, life, and safety response of batteries. Building upon a strong foundation of physics-based models, the emphasis is on delivering practical tools for the broader battery community.

We did not attempt to cover every aspect of the different projects under CAEBAT, but rather provide a flavor of how virtual design tools for large format batteries have evolved under the program. References to earlier publications and webpages are provided throughout the volume for readers to get a comprehensive understanding of the models. Chapter 1 provides an excellent overview of the broad range of computational tools and commercial software that cater to battery engineering. Beyond discussing the capabilities of the various software tools, this chapter also serves as an authoritative compilation of the battery modeling literature. Detailed experimental investigations of heterogeneities in large format cells are discussed in Chap. 2. These one-of-a-kind measurements help us understand thermal and electrical limitations across different parts of the cell. The corresponding models address practical questions that are relevant to scaling up of cell formats. Chapter 3 provides a thorough discussion of electrode engineering that highlight the importance of design at the mesoscale across multiple battery chemistries. These timeless examples showcase the utility of these models for next-generation battery chemistries. Chapter 4 carefully considers aspects of pack design providing a lot of experimental results that call attention to the maturity of these modeling tools and relevance to practical automotive battery design. Chapter 5 covers some of the most detailed experimental studies on parameterizing mechanical response of batteries, from cell components all the way to modules. This chapter calls for attention to detail when setting up abuse tests and provides specific recommendations on minimizing variability in the test results. In Chap. 6 we discuss our experiences simulating mechanical deformation under a variety of abuse conditions. We address some issues frequently called out in this discipline, such as the relationship between contact resistances and heat generation rates or the influence of aging on the abuse response of the cells. Chapter 7 emphasizes multi-scale models and high-performance computing tools to streamline their implementation. The chapter closes out with a detailed discussion of how artificial intelligence and machine learning tools interface with large databanks to redefine computer aided battery engineering. Together, these chapters provide a well-rounded approach to computational design of batteries.

Lastly, this is very much an effort in progress. Alongside newer chemistries, mathematical tools to facilitate battery design and analysis are growing by leaps and bounds. We earnestly hope that the chapters help as much in raising the bar as they do set the stage up for robust design of batteries.

Golden, CO, USA   Shriram Santhanagopalan

Acknowledgments

We acknowledge both technical and financial support from the Vehicle Technologies Office at the Department of Energy's Office of Energy Efficiency and Renewable Energy (EERE) for the Computer Aided Engineering for Electric-Drive Vehicle Batteries (CAEBAT) program under Agreement No. 29834. Special thanks are due to our Program Managers David Howell and Brian Cunningham for the visionary conception of the program to develop a state-of-the-art virtual platform for design and analysis of large format batteries. Funding from U.S. Army TARDEC (now GVSC) under Agreement No. MIPR 11037569/2810215 (Yi Ding) is gratefully acknowledged. Constant support and incessant encouragement from NREL Management, especially, Johney Green, Chris Gearhart, and John Farrell were key enablers in pursuing such a diverse range of projects. Ahmad Pesaran at NREL and John Turner at ORNL were instrumental in assembling, coordinating, and leading the different teams under CAEBAT. Both their years of experience and close work with the relevant partners from Industry were key to the success of the program. Ted Miller from the Ford Motor Company and Mary Fortier from General Motors were early champions of this effort and provided unparalleled access to world-class technical resources. Unwavering support from members of the USABC Technical Advisory Committee, the Crash Safety Work Group, battery companies, and software developers who worked with us providing samples, sharing their technical insights and the needs from the industry deserve our sincere appreciation. We are deeply indebted to our colleagues (current and previous) at NREL for their excellent technical contributions.

Contents

Chapter 1
Applications of Commercial Software for Lithium-Ion Battery Modeling and Simulation

Robert Spotnitz

Abstract Battery modeling on commercial software platforms has its advantages in terms of providing a robust platform for virtual prototyping in realistic geometries, continued support for battery designers, and bridging the knowledge gap across multiple disciplines commonly involved in assembling a battery. This chapter begins with an overview of commercial software engineering tools and expands into specific examples from the literature on popularly used software packages. Where applicable comparisons of capabilities within and features specific to the different tools are provided.

1.1 Introduction

Commercial software for battery design and simulation has flourished since ~2010 being mostly driven by the success of battery electric vehicles. The major applications for commercial battery software are in the design and optimization of systems such as hybrid electric vehicles (HEV) and battery electric vehicles (BEV), battery management systems, and cooling systems. In addition, commercial software is widely used for design of cells, modules, and packs, as well as simulation of aging and abuse. Recently, the use of lithium-ion batteries in grid applications has inspired development of software tools for energy arbitrage using batteries.

Industry uses software as a tool to cost-effectively solve problems. So, the role of modeling and simulation in the lithium-ion battery industry can be appreciated by considering the challenges the battery industry faced over its history. First, some perspective as to the use of simulation for batteries can be obtained by looking at the number of publications dealing with simulation (see Fig. 1.1). Although lithium-ion batteries were commercially introduced in 1991, interest in simulation did not take off until ~2010 and correlates with the growth of the battery electric vehicle market.

R. Spotnitz (✉)
Battery Design LLC, Pleasanton, CA, USA
e-mail: rspotnitz@batdesign.com

S. Santhanagopalan (ed.), *Computer Aided Engineering of Batteries*, Modern Aspects of Electrochemistry 62, https://doi.org/10.1007/978-3-031-17607-4_1

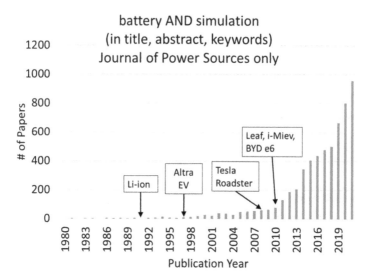

Fig. 1.1 Number of publications indexed by "battery AND simulation" in the Journal of Power Sources from 1980 to 2021. The increase in the number of publications with time is correlated with the introduction of battery electric vehicles

Lithium-ion batteries were first commercially introduced by Sony in 1991 to compete against nickel metal hydride batteries for consumer applications like camcorders. The market demand for higher volumetric energy density (Wh/liter) quickly led to the use of graphite negatives in place of hard carbon, electrolyte additives to improve current efficiency, and later the development of lithium nickel cobalt aluminum oxide positive active materials and then high-voltage cobalt oxide. Also, there were innovations in cell design, especially pouch cells. The invention of lithium iron phosphate led to the widespread use of lithium-ion batteries in power tools as well as industrial applications like forklifts.

Simulation did not play a leading role in the abovementioned advances of lithium-ion batteries because it did not convincingly address the problems of identifying new materials such as electrolyte additives or active materials. Still some software was widely used for predicting lithium-ion cell performance, specifically the physics-based, pseudo-2D (P2D) model of Doyle, Fuller, and Newman [1–3] (DFN) introduced in ~1994. The DFN model proved useful for optimizing cell designs [4].

The DFN model accounted for the major physical processes taking place in a lithium-ion cell. The DFN model accounted for both liquid-phase and solid-phase lithium transport and charge-transfer reactions and electronic resistances. The model successfully explained how liquid-phase mass transport limited discharge capacity and allowed optimization of active material particle size and electrode porosity to maximize energy and power density. These results provide useful guidance for designing lithium-ion cells but are of limited value because of the difficulties in their practical implementation. For example, the DFN model might indicate that a smaller particle size be used to improve rate capability, but changing the particle size also affects the amount of binder and conductivity aid needed to obtain good adhesion and cohesion of the electrode coating and can increase capacity fade over time. The DFN model does not predict this array of side effects from changing particle size, so experimental work is required to evaluate the effect of changing particle size. The DFN model provides useful direction, but industry needs to rely on experimentation to optimize cell designs. Fortunately, the relatively low-capacity cells (<3 Ah) widely used in consumer applications were inexpensive, and industry could rely on extensive testing to optimize designs.

The most urgent problems for the lithium-ion battery industry that addresses the consumer electronics market are increasing energy density (Wh/liter) and lowering cost ($/Wh); safety is, of course, a prerequisite. New materials were thought to be the best path to reach these goals. Simulation provided some guidance for developing new active materials through calculated molecular properties like HOMO and LUMO [5], but the progress is attributed mostly to extensive experimental work and the genius of researchers, notably Professor John Goodenough.

Simulation did aid in the design of power tool batteries because of the importance of tabbing. The placement and number of tabs significantly affected cell resistance, and predictive models were successfully developed [6–8].

The advent of lithium-ion battery electric vehicles (BEVs) created a significant market for software to design and simulate batteries primarily because of the importance of systems design, especially battery management systems (BMS) and battery thermal management systems (BTMS). The battery in an electric vehicle interacts with the electronics, motor, brakes, cooling system, etc., and simulation models can account for these interactions to optimize the overall system. The BMS consists of hardware and software and has many functions, foremost of which is insuring safe operation of the battery [9, 10]. The battery thermal management system includes hardware which provides a mechanism to transfer heat to and from the battery [11].

MATLAB/Simulink dominates the market for vehicle system simulation. Programs such as ADVISOR (https://sourceforge.net/projects/adv-vehicle-sim/) and Autonomie (https://www.autonomie.net/expertise/Autonomie.html) are built on MATLAB/Simulink. Other tools such as Simcenter Amesim and GT-Suite are used as stand-alone tools or to provide specific modules for cosimulation with MATLAB/Simulink of BEVs.

Battery management systems are used in many applications such as aerospace [12] and consumer electronics [13], but the scale of the automotive application incentivized commercial software providers to address the market.

Design of battery thermal management systems usually involve complex geometries and flow of a heat-transfer media like air or liquid coolants. Computer aided engineering (CAE) programs like Simcenter STAR-CCM+ [14] and Fluent [15] were developed to solve computational fluid dynamics (CFD) problems with conjugate heat transfer (CHT), and so are well suited for BTMS design.

The problems of vehicle systems simulation, BMS development, and BTMS design are ongoing as new vehicles are introduced and technology advances. This situation provides a stable or even growing demand for software tools. This dependable demand for battery simulation software supports ongoing maintenance and improvements of commercial software.

1.2 Overview of Commercial Computer Aided Engineering Software

Computer aided engineering (CAE) software is ubiquitous today, being used across most industries in some way. This section provides a brief description of the types of CAE software applicable to batteries to provide some background for the following sections on applications of commercial products.

Electronic design automation (EDA) software like PSPICE has included simple battery models such as a constant voltage source since the 1980s. Over time more complex circuit models were developed; see the recent review of Saldaña et al. [16]. Very sophisticated circuit models that incorporate battery physics are available for use in circuit simulation programs [13]; see Raël et al. [17] for an example using Saber.

System simulation tools like MATLAB/Simulink enable users to construct dynamic models by linking component models of subsystems. For example, a vehicle simulation might include a battery module, a motor module, a passenger compartment module, etc. all linked together.

Multiphysics packages like Simcenter STAR-CCM+ combine the capabilities to mesh complex geometries and apply physics such as fluid mechanics or electromagnetism to produce high-fidelity simulations. These packages are widely used for design of internal combustion engines, ships, rotors, etc. Historically, CAE programs tended to focus either on computational fluid dynamics with conjugate heat transfer using finite volume methods or stress analysis using finite element methods. Over time these packages have included other physics, especially electromagnetism and multi-body dynamics, and optimization tools.

1.3 Commercial Products for Li-Ion Battery Design and Simulation

After 2010, commercial software products for lithium-ion battery design and simulation began to establish markets. The technical foundation for these products was provided by academic research on computer simulation of lithium-ion batteries, especially the overarching work of John Newman [1–3, 18–24]; there are good reviews of academic work before 2010 [25, 26]. Some important work in battery modeling before 2010 was also done by researchers in industry [6]; especially notable is the work of Mark Verbrugge [27–30]. Important work on computer aided engineering of batteries was also done at the National Renewable Energy Laboratory (NREL) led by Ahmad Pesaran [31–36].

In 2010, the US Department of Energy, recognizing the need for computer aided engineering (CAE) tools for the design of electric drive vehicles, initiated a multi-year program led by NREL called Computer Aided Engineering for Electric Drive Vehicle Batteries (CAEBAT). Over the decade since its introduction, the CAEBAT program addressed the following areas (https://www.nrel.gov/transportation/caebat.html):

- MSMD model (CAEBAT-1) – multi-scale multi-domain model developed at NREL [37] for electrothermal simulation of lithium-ion cells
- Software tools for battery cells and packs (CAEBAT-1) – funded tool development at ANSYS, CD-adapco (now Siemens), and EC Power for cell and pack design
- Safety and crush simulation (CAEBAT-2)
- Microstructure applications for battery cells and packs (CAEBAT-3)

Currently there are many companies offering commercial products for lithium-ion battery design and simulation. Here is a partial list of companies and their products:

1. SIEMENS (https://www.plm.automation.siemens.com/global/en/products/)
   In 2007 Siemens acquired the Amesim software which is used for system simulation; the software is used for simulating problems in design of battery electric vehicles such as battery sizing. Siemens acquired CD-adapco and Mentor Graphics in 2016. CD-adapco's flagship product STAR-CCM+ has a batteries module for simulation of cells and packs based using physics-based and empirical models. STAR-CCM+ includes a microstructural model [38] for simulation of electrodes based on reconstructed tomographic models or packings created using a Discrete Element Model. CD-adapco assumed responsibility for the Battery Design Studio® (BDS) in 2012. The BDS software enables sizing and simulation of simple, spiral, and stack lithium-ion cell designs and provides tools for analysis of battery data. Mentor Graphics offers FloTHERM which is useful for design of battery pack cooling systems.
2. Ansys (https://www.ansys.com/products)
   The flagship product Fluent includes battery simulation tools and an application programming interface to the multi-scale multi-domain (MSMD) model developed by NREL [37]. In 2019 Ansys acquired Livermore Software Technology Corporation whose LS-DYNA software is useful for crush simulation of batteries [39].
3. Dassault Systemes (https://www.3ds.com/products-services/)
   Dassault Systemes software offering includes Biovia which include molecular dynamics simulations which have been used for estimation of battery electrolyte properties [40]. The Abaqus software is widely used to study mechanical defor-

mation of batteries; for example, see [41]. The Dymola software is useful for system simulation, and the Catia software is used for electrothermal simulation of cells and packs.

4. Gamma Technologies, LLC (https://www.gtisoft.com/)

    Acquired Autolion® in 2018 which is useful for electrothermal simulation of battery cells. The GT Suite software is used for electrothermal simulation of cells and packs as well as system simulation and battery management systems.

5. COMSOL (https://www.comsol.com/)

    COMSOL offers a Batteries & Fuel Cells Module which is used for simulating electrothermal behavior including runaway, short circuits, and aging. The software allows the use of custom model equations which has made the software extremely popular among academic users.

6. Altair (https://www.altair.com/)

    In 2019 Altair's HyperWorks software added the CellMod™ Virtual Battery by Sendyne which provides a physics-based model for electrothermal simulation of cells and packs.

7. Thermoanalytics (https://www.thermoanalytics.com/battery-thermal-extension)

    Offers a battery extension to their TAITherm package for thermal simulation of cells and packs.

8. Materials Design (https://www.materialsdesign.com/)

    Provide atomic scale modeling of battery materials. Examples include prediction of ionic conductivity and design of active materials.

9. OpenFOAM (https://www.openfoam.com/)

    While OpenFOAM is a free, open-source CFD software, there is commercial support available. There are several examples of the use of OpenFOAM for battery simulation [42–44].

Google Docs hosts a publicly maintained database of battery-related software; see https://docs.google.com/spreadsheets/d/1xBWc2I6vwfTw64-yY_c6iHAMXWgGulekNPf5Y5ntNMM/edit#gid=0.

Mention should be made of the Advanced Electrolyte Model developed by Kevin Gering at Idaho National Laboratory which was at one time offered for license; see [45] for a critical discussion of the model. This software provides a complete set of electrolyte properties [46] which can be used in the Doyle Fuller Newman macrohomogeneous model [1, 3].

1.4 Specific Applications

Visionaries have imagined software workbenches capable of simulating all aspects of batteries from material properties, to manufacturing, to product performance (e.g., see https://www.u-picardie.fr/erc-artistic/). Franco et al. provide an excellent review of multi-scale modeling [47]. However, to-date, commercial software products are directed at specific applications such as BMS or BTMS.

The following sections provide some introduction as to the type of problems commercial software addresses in the designated application area. The literature surveyed in each section is not comprehensive but focusses on some major software products. The intent is to provide the user with an overview of the type of problems the major software packages address.

1.4.1   Materials Design

The capabilities of modeling to aid in materials design are discussed in recent reviews. Urban et al. [48] provide an overview of state-of-the-art ab initio approaches for the modeling of battery materials and Wang et al. [49] review modeling of the anode SEI. Wikipedia provides lists of software packages for quantum chemistry (https://en.wikipedia.org/wiki/List_of_quantum_chemistry_ and_solid-state_physics_software) and molecular mechanics (https://en.wikipedia. org/wiki/Comparison_of_software_for_molecular_mechanics_modeling). Here are some applications of commercial software related to battery materials.

Atomistic modeling of battery materials tends to be carried out by academics using public domain software. However, Materials Design Corporation does provide some examples (see https://www.materialsdesign.com/batteries). Also Dassault Systems has demonstrated use of their Biovia software for electrolyte property estimation [40].

The DUALFOIL model [3] incorporated concentrated solution theory; however, there were virtually no complete sets of property data available to use in the model. Valoen and Reimers [50], working at E-One Moli Energy (Canada) Limited, experimentally measured the complete set of transport properties (density, transport number, activity coefficient, diffusion coefficient, and conductivity) and demonstrated that the DUALFOIL model predicted experimental results. Later Gering [51, 52] developed the Advanced Electrolyte Model for predicting complete property sets for electrolytes. A number of these electrolyte property sets are included in the Battery Design Studio software [46]. Logan and Dahn [45] provide a review of approaches for determining electrolyte properties.

Some workers have used ANSYS software for simulation of mechanical behavior of materials. For example, Ramasubramanian et al. [53] used ANSYS Mechanical software to represent a coupled diffusion and structure mechanics approach for simulation of lithiation of tin oxide nanowires. Tokur et al. [54] modeled mechanical properties of various silicon composite structures.

Clancy and Rohan [55] used Comsol to compare the performance of active materials in different geometries, solid-state thin-film micro, nanowire and core-shell nanowire. Painter et al. [56] developed a 3D phase field model in Comsol for lithiation of a $LiFePO_4$ particle.

1.4.2   Electrode Simulation

Microstructural models account for the geometry of the active material particles in the electrodes and have been developed for use in Siemens Simcenter STAR-CCM+ [57], Ansys Fluent [58], and Comsol; Zhang et al. have recently reviewed microstructural models [59]. Microstructural models use the theory developed by Fuller et al. [3] but eliminate the assumption of macro-homogeneity by resolving the microstructure so can be applied to a geometry obtained from 3D reconstruction from X-ray tomography, focused ion beam/scanning electron microscopy, or the discrete element method. A microstructural model can be used to quantify the tortuosity of an electrode and account for its variation in different directions [60]. Microstructural simulations are computationally intensive and applied to small regions, for example, ~100 μm x 100 μm x 100 μm. Below are examples of the use of commercial software for microstructural simulations.

Wiedemann et al. [58] used a microstructural model (Ansys) to explore spatial variations of discharge behavior within a lithium cobalt oxide electrode structure resolved by FIB-SEM. Goldin et al. used 3D microstructure model (Ansys) to assist parameterization of 1D model [61]. ChiuHuang and Huang [62] developed a thermal-static coupling finite element model (Ansys) to study the diffusion-induced stresses under various C-rate conditions on lithium iron phosphate electrodes.

Hutzenlaub et al. [38] resolved the carbon-binder phase in a lithium cobalt oxide electrode and simulated the behavior using a microstructural model (STAR-CCM+, Siemens). Cooper et al. [60] used STAR-CCM+ (Siemens) to compute the directional tortuosity of an iron phosphate electrode.

Higa et al. [63] compared a macrohomogeneous model (Comsol) of an NMC electrode to a microstructure model (STAR-CCM+, Siemens).

Wang [64] used Comsol to develop a microstructure model of a silicon electrode that accounted for chemo-mechanical coupling.

1.4.3   Cell Design and Simulation

The problem of cell design involves specification of the dimensions and composition of components such as foils used for current collectors and tabs, electrode coatings, separator, electrolyte, miscellaneous internals such as tape and insulators, and packaging. This problem is often carried out by battery companies using spreadsheets. Many commercial software products, with the notable exception of Battery Design Studio, take a cell design as an input and provide capabilities for modeling the performance of that cell.

Usually, the first step in cell design is cell sizing. The jellyroll or stack must be sized to fill some specified volume. For spirally wound cells, the assumption of an Archimedean spiral is often made but may not necessarily represent the geometry of cells with one or more tabs placed in the winding. The Battery Design

Studio software includes a winding simulation that accounts for many electrode designs used commercially. Given the lengths of electrodes and separators, the winding simulation predicts the jellyroll size to ensure that it will fit in a specified diameter can. Alternatively, given a target diameter, the lengths of the electrodes and separators can be computed. Battery Design Studio automatically transfers the results of the winding simulation to the model to avoid any errors.

An error made by some users when designing spiral cells is to only consider the coated electrode area on one side of the electrode and so enter a value for electrode area that is half of the correct value. Users sometimes do not study the jellyroll geometry to verify that the design is feasible. For example, when a graphite negative is used, the graphite coating should extend beyond the positive coating to avoid lithium deposition. Thus, the capability to visualize the spiral and inspect the positions of the coatings and tabs is important.

The early work of Harb and LaFollette accounted [65] for current flow in lead-acid spiral cells. For lithium-ion cells, Reimers [6, 66] developed a method for accounting current flow in spiral cells. Lee et al. [7] modeled spiral cells as a concentric cylinders. Spotnitz et al. [8] also developed a method for simulating current flow in spiral cells which did not rely on the assumption of an Archimedean spiral and is useful for both cylindrical and prismatic winds.

Sizing of stack cells is relatively straightforward. To maximize the use of the available volume available, the number of electrodes and the thickness of the positive and negative coatings can be adjusted; normally, the ratio of thickness of the positive coating to the thickness of the negative coating is kept constant.

As described above, the Battery Design Studio (Siemens) software focuses on the problem of cell design. Essentially starting from a blank sheet, the user can virtually construct a cell and estimate performance. This approach enables the potential benefits of new materials to be estimated; for example, Howard and Spotnitz [67] evaluated the theoretical energy density of 18650-size cells with different metal phosphate positives. However, rather than design, the problems of modeling and simulating the performance of an existing lithium-ion cell have received more attention in the literature.

Simulating the performance of a given design is typically done with an empirical model such as NTGK (Newman Tiedemann Gu Kim) [68, 69] or RCR [70, 71] or a physics-based model such as DFN. With empirical models, one should be wary of model predictions outside of the fitted parameter space. Physics-based models are more computationally intensive than empirical models and can be implemented in different ways. For example, the DFN model can be implemented as P2D, P3D, and P4D [66]. Battery Design Studio applies a P2D implementation of the DFN model at each node of the positive electrode so does not account for lithium transport to regions of the negative electrode that overlap the positive. To account for edge effects would require a P3D or P4D model which would be significantly more computationally expensive. This point can be illustrated using Fig. 1.2 which illustrates a discretization of an electrochemical cell.

Referring to Fig. 1.2, if the electrochemical element e is a P2D model, then the y-direction is one dimension, and the second pseudo-dimension is the particle

Fig. 1.2 Schematic of
discretized electrochemical
cell. The current distribution
in the positive (+) and
negative (−) plates can be
solved using Poisson's
equation, while the
electrochemistry element
between plates (e) can be an
empirical or physics-based
model

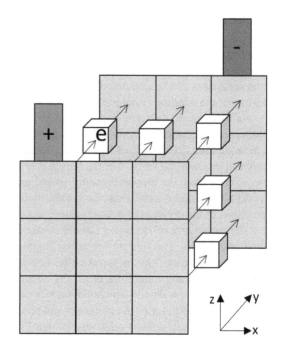

radius. If the electrochemical element is a P3D model, then y/x or y/z are two of
the dimensions, and the third pseudo-dimension is the particle radius. Finally, if the
electrochemical element is a P4D model, then y, x, and z are three of the dimensions,
and the fourth pseudo-dimension is the particle radius. Figure 1.2 indicates there are
nine electrochemical elements, one for each area segment of the current collector.
A common approximation is to use one electrochemical element for the entire
plate. When one electrochemical element is used to represent the entire plate, the
variation in current and voltage across the plate can still be accounted for using
Poisson's equation. However, if the variation in current and voltage across the plate
is not considered, then there may be significant error in the computed temperature
distribution.

Referring to Fig. 1.3, the current to the cell flows into the negative electrode
from the negative tab and exits the cell through the positive tab. The current in the
negative electrode is at its highest value near the tab and at its lowest value at the
bottom end. The ohmic heat is proportional to the square of the current, and the
ohmic heat is highest near the tab and lowest at the bottom of the cell. If the current
distribution in the current collector is not accounted for, then the heat generation will
not be accurately distribution which may produce a significant error in the computed
temperature distribution. Despite this risk, the current distribution in the current
collectors is often not considered because the user does not know the design of the
current collection or has concern about the computational cost.

CAE packages like STAR-CCM+ or Fluent typically generate more complex
meshes than the simple rectangular grid shown in Fig. 1.2. CAE packages use

Fig. 1.3 Current flow
through a battery. Current
flows into the negative tab
into the negative electrode
(black). The current in the
negative diminishes as it
flows downward because
some is transferred to the
positive electrode (red). The
total current is applied to the
negative electrode near the
tab, and the current flow in
the negative is zero at the
bottom of the negative
electrode; the same situation
applies to the positive
electrode

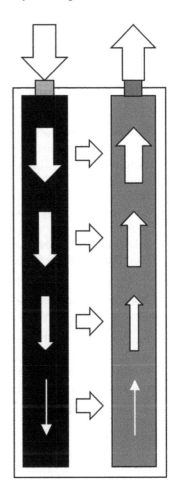

volume elements based on shapes such as polyhedrals, tetrahedrals, or hexahedrals. These meshes usually involve a much larger number of volume elements than simple prismatic meshes, especially if each region (collector, coating, separator) must be resolved. The same mesh is used for the electrochemistry, and thermal problems can be computationally expensive unless simple models like NTGK or RCR are used to represent the electrochemistry. Kim et al. [37] developed the multi-scale multi-domain (MSMD) approach which eliminates the need to resolve the individual layers; LS-DYNA offers Tshell elements that encompass five layers into a single element [72]. Still, when the number of volume elements is ≥1000, simpler models such as the NTGK or RCR, or even a reduced order DFN model, are preferred to reduce the computational cost.

Rather than use the same mesh for the electrochemistry and thermal problems, the heat generation from the electrochemistry problem can be mapped onto the mesh for the thermal problem. This approach allows use of the full DFN model at reasonable computational cost. Many of these different options for the electrochemical and thermal problems are provided by commercial software.

The usage of programs provided by four major providers (Siemens, Ansys, Comsol, Gamma Technologies) is surveyed below.

There are several publications that involve the use of Battery Design Studio (Siemens):

- Howard and Spotnitz [67] evaluated the theoretical energy density of 18650-size cells with different metal phosphate positives.
- Yeduvaka et al. [73] developed models for Li/MnO$_2$ coin cells from several manufacturers.
- Hartridge et al. [74] developed a model for a lithium-ion pouch cell.
- Spotnitz et al. [46] explored the effect of electrolyte composition on cell performance.
- Sakti et al. [75] validated the capability to simulate the discharge behavior 18650-size NMC/graphite cells based on cell dissection.
- Bandla [76] simulated a NMC811/graphite-SiOx cell.

A number of publications describe electrothermal and mechanical modeling of lithium-ion cells with Ansys software; for example:

- Panchal et al. [77] developed an electrothermal model (MSMD) for an 18650-size lithium-ion cell to study the discharge behavior.
- Panchal et al. [78] developed an electrothermal model (MSMD) for a 20 Ah LiFePO4/graphite pouch cell.
- Vyroubal et al. [79] tested a pouch cell at different rates using a thermographic camera to characterize the temperature and compared the results to an electrothermal model (MSMD).
- Oh and Epureanu [80] used the structural analysis package of Ansys to predict the shape of a prismatic cell due to thermal swelling and intercalation-induced expansion.
- Kleiner et al. [81] modeled (MSMD with RCR circuit) a single 25 Ah NMC/graphite prismatic cell with boundary conditions that approximate an actual pack.
- Li et al. [82] used electrothermal model (MSMD with P2D DFN) to study electrothermal behavior of a 10 Ah pouch cell with LiFePO4/graphene hybrid cathode and graphite anode.
- Li et al. [83] used a physics-based (MSMD P2D DFN) electrothermal model to simulate a drive cycle for a 14.6 Ah prismatic lithium-ion battery.
- Li et al. [84] modeled the current interrupt device and vent mechanism of an 18650-size cell.
- Pan et al. [85] used LS-DYNA to simulate the deformation of a 17 Wh, LCO/graphite pouch cell under UN 38.3:2015 T.6 impact/crush test.
- Deng et al. [86] used LS-DYNA to simulate the impact of semi-cylindrical and semi-spherical indenters on lithium-ion pouch and prismatic cells.

- Sheikh et al. [87] used LS-DYNA to simulate compression (rod, circular punch, 3-point bend, flat plate) of 18650-size LiCoO2/graphite cells.
- Xi et al. [88] used LS-DYNA to explore the effect of impact velocity on failure of a cylindrical NMC/graphite cell due to both radial and 3-point bending from 1 to 20 m/s.

Comsol has also been widely used for simulating lithium-ion cells, for example:

- Somasundaram et al. [89] used a DFN model to study the electrothermal behavior of a 18650-size $LiMn_2O_4$/graphite cell with both a convective boundary condition and a phase-change material.
- Xu et al. [90] experimentally verified a 2D electrothermal DFN model for a cylindrical 38120 $LiFePO_4$/graphite cell.
- Samba et al. [91] used an experimentally validated 3D electrothermal model to compare the effect of tab locations on a 45 Ah $LiFePO_4$/graphite pouch cell.
- Falconi [92] and Frohlich [93] studied the parameter space of the DFN model with an eye toward electric vehicle applications.
- Cai and White [94] used Comsol's Newman model to explore the effect of different cooling conditions on cell temperature during discharge.
- Hosseinzadeh et al. [95] used the DFN model to optimize a unit cell sandwich of $LiFePO_4$/graphite for specific energy and power.
- Dai et al. [96] evaluated the discharge behavior of different cathode materials in a 2D U-shaped battery with a lithium anode.
- Gao et al. [97] developed a DFN model with an aging mechanism and used Comsol to validate the model.
- Bizeray et al. [98] solved the DFN model equations using orthogonal collocation and used Comsol to validate their results.
- Farrag et al. [99] used Comsol model to optimize the charging rate of a lithium-ion battery.
- Singh et al. [100, 101] used the DFN model to explore effect of magnetic fields on performance.
- Painter et al. [102] generalized a single particle model to include an energy balance.
- Rajabloo et al. [103] used MATLAB to carry out a parameter estimation of Comsol's single particle model.
- Lam and Darling [104] implemented a method in Comsol to switch between constant current and constant voltage modes of the DFN model.
- Tang et al. [105] used a three-dimensional electrothermal model to study the solid-liquid potential variation across the cell.
- Li et al. [106] modeled the spiral of an 18650-size cell in two dimensions using a Newman model for the electrochemistry. The effect of the tab design on the electrothermal behavior was studied.
- Chiew et al. [107] developed a pseudo-3D electrothermal model of a 26650-size $LiFePO_4$/graphite cell. The model fits the temperature of the cell over a range of ambient temperatures and discharge rates.

- Tan et al. [108] studied the effect of initial temperature on charging of an 32650-size LiFePO$_4$/graphite cell.
- Li et al. [109] developed a coupled mechanical-electrochemical-thermal model to study the short-circuit mechanism of a lithium-ion pouch cell due to mechanical indentation.

The Autolion software has been used to for a variety of cell-level simulations, for example:

- Sripad and Viswanathan [110] evaluated different lithium-ion battery chemistries for electric vehicles.
- Dareini [111] demonstrated how to parameterize a battery model of a 20 Ah pouch cell.
- Tanim [112] used Autolion to evaluate the accuracy of a single particle model.
- Wang et al. [113] used Autolion with Ansys Fluent to simulate nail penetration.
- Yang et al. [114] describe a model for lithium metal plating that can fit the capacity fade behavior of a NMC622/Graphite cell under different charge rates.

Mention should be made of the extensive use of Abaqus for modeling mechanical deformation of lithium-ion cells [41, 115–118].

Though not a commercial product, OpenFOAM can be used to model the electrothermal behavior of a prismatic lithium-ion cell (see Darcovich et al. [44]).

To provide some appreciation of using a tool like Battery Design Studio (BDS), the paper "A validation study of lithium-ion cell constant c-rate discharge with Battery Design Studio®" by Sakti et al. [75] is examined in some detail here. The first step is to define the physical construction of the cell. The authors procured some commercial NMC111/graphite, 2.05 Ah 18650-size cells, and dissected a cell to obtain the cell geometric details. BDS provides a structured list to define the cell components:

- Positive electrode

    - Formulation

        Active materials
        Binders
        Additives

    - Tab
    - Tape
    - Collector

- Negative electrode

    - Formulation

        Active materials
        Binders
        Additives

    - Tab

- Tape
- Collector

- Separator
- Electrolyte
- Package
- Internals (e.g., plastic insulators)

The dimensions of the electrodes, separator, and package are obtained from the dissection. Reasonable estimates of the formulation were made, and equilibrium voltage curves for the active materials were taken from the literature. BDS allows voltage curves to be input as tables and automatically fit to a monotonic cubic spline.

Once the components have been defined, the cell is "built" using a winding simulation which computes the active area of the cell and generates a report detailing the theoretical cell capacity, weight, volume, and energy density. Some fine-tuning of the cell design such as adjusting the loadings of the coatings can be done to make sure that the cell weight and capacity agree with the actual values.

Once the values in the report agree with the actual values, a simulation model can be selected. The authors chose the P2D DFN model and used default values for the exchange current densities and solid-phase diffusion coefficients.

Finally, simulations were carried out at C/5, C/2, 1C, and 2C discharge rates and voltage versus capacity curves compared to those measured by the authors and those reported by the manufacturer. The discrepancy between BDS and the manufacturers data was less than 5% for the total energy delivered. While these results were satisfactory, more generally all the parameters would have to be measured (e.g., see Ecker et al. [119, 120]) and the current collection option of BDS selected to account for the voltage drops in the tabs and current collectors.

1.4.4 Module and Pack Design

Making modules or packs of cells introduces the concepts of an electric circuit where cells are connected in series and/or parallel and a structure for heat exchange (see Santhanagopalan et al. [121] for an introduction to pack design). Deng et al. [122] review modeling of lithium-ion batteries, including packs, under abuse conditions.

As mentioned earlier, commercial CAE software is well-suited to the problem of analyzing battery pack designs because of its capabilities to mesh complex geometries and simulate fluid flow and heat transfer.

Examples from the major software providers (Siemens, Ansys, Comsol, and Gamma Technologies) are provided.

Here are some examples of using Siemens software for simulating battery packs.

- Kawai [123] used STAR-CCM+ to study multifunctional spacers for preventing propagation of thermal runaway.
- Basu et al. [124] used STAR-CCM+ to evaluate a novel liquid-cooling system for 18650 cells.
- Deng et al. [125] used STAR-CCM+ to study a module consisting of 10 Ah prismatic LiFePO4/graphite cells with liquid-cooled plates between cells. The effects of the coolant flow rate, number of plates, channel distribution, and cooling direction were studied.
- Deng et al. [126] used STAR-CCM+ to do a simulation study of a cooling plate with a novel leaf-like channel structure used to cool 45 Ah prismatic cell. Deng et al. [127] presented an optimal design for cooling plates with bifurcating mini-channels in an 8 Ah prismatic cell module.
- Sheng et al. [128] used FloEFD to simulate liquid-flow in serpentine channels of a cooling plate used between prismatic cells.
- Lajunen and Kalttonen [129] used AMESim to simulate an electric city bus over different drive cycles.
- Tran et al. [130, 131] used AMESim to evaluate the use of a heat-pipe to cool hybrid and battery electric vehicle modules.
- Wang et al. [132] used AMESim to develop an electrothermal model for an air-cooled module consisting of 12 NMC cells in series.
- Wang et al. [133] used AMESim to develop an electrothermal model for an liquid-cooled PHEV battery (1P84S) consisting of seven modules using 37 Ah NMC cells.

There are several examples of using Autolion to simulate battery performance.

- Vora [134] used Autolion to evaluate the effect of aging on HEV powertrains.
- Berglund [135] used GTSuite to evaluate the effect of driving style on energy usage in BEVs.
- Cheng et al. [136] used Autolion to explore the effect of temperature on aging of HEV batteries under a standard driving cycle.
- Xing et al. [137] compare different aging models, including Autolion, for system design and development of control algorithms.
- Carnovale [138] parameterized the aging models in Autolion to study the effect of thermal management.

There are numerous examples of using Ansys software to simulate battery performance:

- Li et al. [139] used Fluent to simulate the thermal behavior of a square-packed module of eight 26650-size LiFePO4 cells (2P4S configuration) in a wind tunnel.
- Liu et al. [140] developed an electrothermal model (MSMD with NTGK) for a lithium-ion SLI battery. The battery consisted of four parallel-connected modules where each module consisted of four series connected 20 Ah NMC/Graphite pouch cells.

- Sun et al. [141] used Fluent to model air flow through ducts and channels of a module consisting of prismatic cells spaced by air channels.
- Xun et al. [142] evaluated the effect of cooling channel size on stacks of both cylindrical and prismatic cells.
- He et al. [143] modeled the air flow through an eight-cell module of 26650-size LiFePO$_4$/graphite cells.
- Panchal et al. [144] used Fluent to develop a model for a minichannel cold plate used with a water-cooled LiFePO$_4$ battery.
- Wang et al. [145] used Fluent to evaluate five different cooling schemes ((1) air natural convection, (2) natural convection with fins, (3) forced air convection with fins, (4) forced convection with slotted fins, and (5) natural convection with fins and PCM) for a module of pouch cells.
- Tang et al. [146] used Fluent to simulate the cooling of six 50 Ah LiFePO$_4$/graphite prismatic cells. Different cooling plate arrangements (bottom, sides, bottom + sides) were evaluated.
- Zhao et al. [147, 148] used Fluent to simulate a battery module consisting of 71 18650-type NMC/graphite lithium-ion cells cooled by channeled liquid flow. The module design is similar to that used by Tesla where the cooling channel follows a serpentine path through the array of cells and makes a U-turn, so the inlet and outlet are on the same side. Xie et al. [149] carried out a similar study and recommend adding baffles in the micro-channel to improve the cooling.
- Li et al. [150] used Fluent to simulate a pack of 14 pouch cells containing cooling plates.
- Li et al. [151] used Fluent to simulate an air-cooled module of eight cells; these simulation results were used to generate a surrogate model for an optimization study to minimize pack volume, minimum temperature difference, and standard deviation of minimum temperatures.
- Zhou et al. [152] used Fluent to design a novel air cooling system for a pack of 18650-size NCM/graphite cells where perforated tubes were placed in the interstitial space formed by a square packing of the 18650-size cells.
- Wang et al. [153] used Fluent to evaluate air cooling of a pack of 18650-size cells. The effect of the cell arrangement and air flow inlet and outlet positions was studied.
- Chen et al. [154] used Fluent to optimize the channels for a liquid-cooled plate stacked between 8 Ah pouch cells.
- Chen et al. [155] used Fluent to evaluate a number of duct designs to feed air channels between prismatic cells.
- Yang et al. [156] used Fluent to evaluate the effect of axial air cooling on a module of 26650-size LiFePO$_4$/graphite cells.
- Na et al. [157] used Fluent to study the thermal behavior of modules of cylindrical cells partitioned so that air can reverse flow across the cells.
- Wang et al. [158] used Fluent to optimize a liquid-cooled module consisting of 90 18650-size NCM/graphite cells in 6 rows of 15 cells with plates between rows. The effect of inlet velocity, channel number, and contact angle on the cooling performance was studied.

- Jiaqiang et al. [159] used Fluent to model liquid flow through channels of a pack of prismatic cells.
- Song et al. [160] used Fluent to simulate the thermal behavior of a module of 18650-size cells embedded in phase-change material and sitting on a liquid-cooled plate.
- Liu and Zhang [161] used Fluent to simulate the air flow through a stack of prismatic cells with variable air gap spacing across the module.
- Cao et al. [162] used Fluent to simulate liquid flow through channels in a full-size-scale cylindrical battery pack containing 5664 18650-size lithium-ion cells.
- Yang et al. [163] used Fluent to demonstrate the benefit of using liquid metal such as gallium as a coolant for lithium-ion batteries.
- Chung and Kim [164] used Fluent to model the thermal behavior of a module of prismatic cells with metal fins between the cells that connect to a bottom cooling plate that the cells sit upon.
- Lai et al. [165] used Fluent to evaluate a cooling channel for a module of cylindrical 18650-size cells where the coolant channels serves as a lightweight spacer for the cells.
- Zhang et al. [166] used Fluent to carry out electrothermal simulations of a battery module comprised of pouch cells in a 6S4P configuration with coolant channels spaced between them.
- Zhou et al. [167] used Fluent to model the thermal behavior of a module of 18650 cells where the cells are helically wrapped with a duct.

Here are several examples of using Comsol to simulate battery modules:

- Kizilel et al. [168] simulated the thermal behavior of small cylindrical cells housed in a phase-change material (PCM) using Comsol.
- Li et al. [169] used Comsol to simulate air cooling of a battery module with silica enclosed copper mesh plates.
- Yang et al. [170] used Comsol to compare the electrothermal behavior of a module of cylindrical cells in a square versus a staggered configuration cooled by air flow.
- Mohammed et al. [171] used Comsol to study different designs of water-cooled plates with 20 Ah prismatic LiFePO4/graphite cells under normal conditions and under thermal runaway.
- Al-Zareer et al. [172] used Comsol to evaluate a novel cooling scheme based on vaporization of liquid propane in a pack of 18650-size LiMn2O4/graphite cells.
- Bugryniec et al. [173] simulated propagation of thermal runaway in a pack containing 9 cylindrical 18650 LiFePO4/graphite cells.

Though not a commercial product, OpenFOAM can be used for pack design. For example, Darcovich et al. [43] used OpenFOAM to compare the effectiveness of cooling plates flush with cell face versus on the bottom surface of cell.

To provide some appreciation of using a tool like STAR-CCM+, the paper of Basu et al. [124] "Coupled electrochemical thermal modelling of a novel Li-ion battery pack thermal management system" is described in more depth. The battery

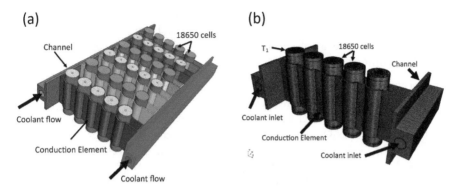

Fig. 1.4 Geometry of the pack and cooling system (**a**) and the mesh generated in STAR-CCM+ (**b**). (Printed with permission from [124]. Copyright 2016, Elsevier)

pack consists of 30 NCA/graphite, 18650-size cells (6S5P) with water coolant channels on the sides and aluminum heat conduction elements to support the cells and transfer heat to the coolant channels; the geometry of the battery pack is shown in Fig. 1.4a and b shows the mesh for the first row of parallel cells comprised of 400,000 polyhedral cells. Battery Design Studio (BDS) was used to generate a model for the 18650-size cell; the discretization of the current collectors was considered with P2D DFN model to represent the electrochemistry. STAR-CCM+ constructed the geometry for the 18650-size cells from the BDS input, and the user supplied CAD for the coolant channels and heat conduction elements. STAR-CCM+ generated a mapping between the electrical mesh with the electrochemical elements (see Fig. 1.2) and the polyhedral mesh (Fig. 1.4b) used for flow and heat-transfer. Experiments were done to validate that the model accurately predicts the temperature of the cells. Then the model was used to predict the effects of coolant flow rate, contact resistance between the cells and heat conduction elements, and discharge rate. These simulations provide confidence that the battery thermal management system is viable.

1.4.5 Battery Management Systems

Battery management systems (BMS) control and monitor batteries for charging, safe operation, cell balancing, and determining state of charge, state of health, etc. [174]. The software use in a BMS is usually customized for specific batteries. As mentioned earlier MATLAB/Simulink dominates the market for BMS development because the tool is designed for code development. The BMS literature consists mainly of methods for obtaining specific metrics rather than providing examples of commercial software tools. For example, see Hannan et al. [175] and Ali et al.

[176] for reviews of state-of-charge estimation methods. However, there are some examples of using commercial software for BMS development.

Here only some examples using Siemens software are provided as there is no significant difference in usage of other products:

- Liu et al. [177] describe a AMESim/Simulink cosimulation for real-time prediction of the SOC of a lithium-ion battery.
- Chen et al. [178] describe a AMESim/Simulink cosimulation for real-time prediction of the SOC and SOH of a lithium-ion battery.
- Locorotondo et al. [179] used Amesim to evaluate a SOH algorithm using a model for a LiFePO4 cell with aging.
- Veeraraghavan et al. [180] developed a BMS to estimate capacity using AMESim to simulate the powertrain.
- Lee et al. [181] used Amesim to develop a design methodology for electric vehicles based on optimization of reliability and market properties (such as profit or social welfare).

1.4.6 System Design

For simulation of how the battery interacts with the vehicle, a system simulation tool like MATLAB/Simulink, Amesim, GT-Suite, etc. is used. The approach is typically to use 0D or 1D models though 3D model can be used for detailed simulations of specific components.

Mahmud and Town [182] provide a review of computer tools used to model electric vehicle requirements that covers some 44 different packages. Ali et al. [183] review tools for energy management of hybrid electric vehicles. Ringkjøb et al. [184] review some 75 modeling tools for energy systems.

Wang et al. [185] review approaches for simulation of vehicle-integrated thermal management which accounts for the thermal and mass flow distribution of a complete vehicle (e.g., including the air conditioning system).

MATLAB/Simulink dominates this market because it serves as a foundation for other products; see examples below. Some examples from Siemens software are provided; Gamma Technologies software is used in a similar way.

1.4.6.1 MATLAB/Simulink

- The Advisor® software (https://sourceforge.net/projects/adv-vehicle-sim/) is "MATLAB/Simulink based simulation program for rapid analysis of the performance and fuel economy of light and heavy-duty vehicles with conventional (gasoline/diesel), hybrid-electric, full-electric, and fuel cell powertrains." There are hundreds of articles that use the open-source Advisor software to simulate vehicles.

- Autonomie (https://www.autonomie.net/expertise/Autonomie.html) is a very widely used MATLAB-based system simulation tool for vehicle energy consumption and performance analysis. Applications include energy consumption and performance analysis throughout the entire vehicle development process (e.g., MIL, HIL, SIL).
- Park et al. [186] developed a model for an electric vehicle powertrain and simulated a drive cycle to analyze the effect of the various components.
- Liu et al. [187] used MATLAB/Simulink to minimize the life cycle cost of a hybrid electric energy storage system (battery and DC/DC convertor) for construction and mining vehicles with dynamic on/off loads.
- McWhirter et al. [188] developed a MATLAB model to minimize the energy consumption of a series hybrid system, the Bradley fighting vehicle.

1.4.6.2   Siemens

Here are some examples of system simulation which use Siemens software:

- Broglia et al. [189] used AMESim to model an electric vehicle to determine the effect of passenger cabin temperature on range.
- Badin et al. [190] used AMESim to evaluate the effect of driving conditions, auxiliaries use, and driver's aggressiveness on energy consumption of battery electric vehicles.
- Irimia et al. [191] used AMESim to simulate an electric vehicle based on the Renault ZOE with a validated battery model.
- Vatanparvar et al. [192] used AMESim to simulate energy consumption of battery electric vehicles in order to find driving routes that minimize energy consumption.
- Faruque and Vatanparvar [193] used AMEsim to model an electric vehicle to assess the impact of the heating, ventilation, and air conditioning (HVAC) system on energy consumption.
- Zhang et al. [194] used an AMESim/Simulink cosimulation to model the hydraulic components of the regenerative braking system.
- Daowei et al. [195] build an AMESim model to explore the effect of upper and lower SOC limits on fuel economy of a parallel HEV.
- Husar et al. [196] use AMESim to compare functional and structural approaches for simulation of an EV. Both approaches give comparable results though the functional approach is more computationally efficient.
- Berzi et al. [197] developed an Amesim/Simulink cosimulation to evaluate various regenerative braking approaches.
- Li et al. [198] describe development of a vehicle level simulation in Amesim useful for BMS, hardware-in-the-loop, and software-in-the-loop simulation. Starting from experimental data for a nickel-rich, silicon-graphite 18650-type lithium-ion cell, the parameters for a circuit model are obtained. The cell model

is used to construct a module (20P12S) and then a pack of 16 modules in series, and the pack is used in a vehicle model with a BMS.

To illustrate the utility of system simulations, the work of Li et al. [198] is examined. Figure 1.5 shows the vehicle system model generated by Li et al. [198]. A drive cycle (WLTC, Worldwide harmonized Light-duty vehicles Test Cycles) was applied to the vehicle, so the vehicle model could compute the required load on the batteries and the battery model compute the heat generation. The effect of cooling the pack was explored by turning on air cooling when 1 of the 48 battery models reached 35 °C. The battery pack model is computationally efficient that it can be run in real-time for SiL and HiL tests.

Very recently, LS-DYNA has been used to simulate the full-vehicle crash simulations with a battery [199].

1.5   Conclusions

First some specific comments for each of the major application areas.

1.5.1   Materials Design

The use of commercial software for design of active materials and electrolytes has limited application. The capability to predict electrolyte properties (diffusivity, conductivity, transport number, activity coefficient, density) seems within reach. Atomistic simulations to predict solid-phase diffusion coefficients of actual particles, equilibrium curves, interfacial stability, etc. are mostly of academic interest. Engineering approaches to design of active material particles are possible [200] but not yet advocated by software providers.

1.5.2   Electrode Design

Several commercial packages offer microstructural models, but the use of these models does not seem to be widespread. Microstructural models are computationally intensive, and the large volume of results is not easily comprehended. More basic work like the ARTISTIC project (see https://www.u-picardie.fr/erc-artistic/) is needed to generate models that predict how formulation and processing of electrode coatings determine their geometry. Such work would enable optimization of electrode coatings.

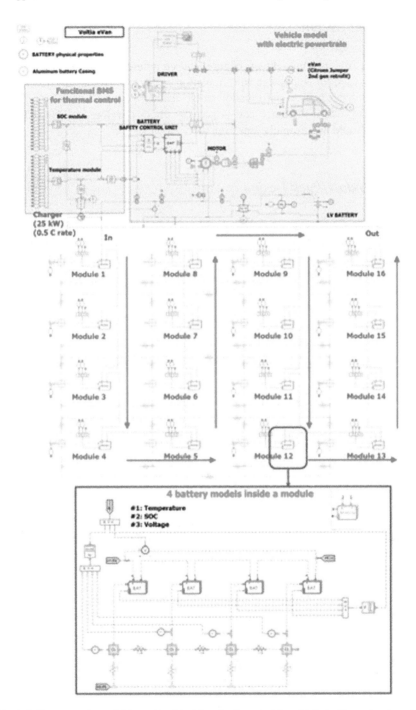

Fig. 1.5 Battery pack model in a virtual test bench of an electric vehicle in Simcenter Amesim. (Printed with permission from [198]. Copyright 2020, MDPI)

1.5.3   Cell Design

Designs for lithium-ion cells are mature, and commercial software programs are now widely used for cell design and simulation of electrothermal behavior. Adding simulation of mechanical deformation to electrothermal simulations is important for predicting abuse tolerance and is actively under development [84, 85]; this capability should make simulations even more valuable.

1.5.4   Module/Pack Design

Design of battery thermal management systems is an established application of commercial software. Simulation can provide excellent fidelity to actual performance. Commercial software can directly consider effects such as thermal expansion and even the expansion of pouch cells due to lithiation. As in cell design, adding the capability to simulate mechanical deformation to predict abuse tolerance will make this software even more essential.

1.5.5   Battery Management and System Design

Use of system design software for battery systems is well established, but the growth of the battery electric vehicle market has stimulated commercial vendors such as Amesim and GT-Suite to dedicate more attention to this area.

Commercial software for battery simulation has found a significant market thanks to the success of the battery electric vehicle market. The application of commercial software is expanding and its functionality increasing to address more problems such as abuse. There is now a positive feedback loop between software providers and the battery industry wherein improvements in software lead to improvements in the battery industry which support further improvements in software and so on. Consequently, we can expect an expanding role for commercial software in battery simulation.

References

1. Doyle M, Fuller TF, Newman J (1993) J Electrochem Soc 140:1526–1533
2. Fuller TF, Doyle M, Newman J (1994) J Electrochem Soc 141:962–990
3. Fuller TF, Doyle M, Newman J (1994) J Electrochem Soc 141:1–10
4. Doyle M, Fuentes Y (2003) J Electrochem Soc 150:A706–A713
5. Yuan L-X, Wang Z-H, Zhang W-X, Hu X-L, Chen J-T, Huang Y-H, Goodenough JB (2011) Energy Environ Sci 4:269–284

6. Reimers JN (2006) J Power Sources 158:663–672
7. Lee K-J, Smith K, Pesaran A, Kim G-H (2013) J Power Sources 241:20–32
8. Spotnitz R, Hartridge S, Damblanc G, Yeduvaka G, Schad D, Gudimetla V, Votteler J, Poole G, Lueth C, Walchshofer C, Oxenham E (2013) ECS Trans 50:209–218
9. Xing Y, Ma WWM, Tsui KL, Pech M (2011) Energies 4:1840–1857
10. Plett GL (2016) Battery management systems, vol II. Artech House, Norwood
11. Kim J, Oh J, Lee H (2019) Appl Therm Eng 149:192–212
12. Damiano A, Porru M, Salimbeni A, Serpi A, Castiglia V, Tommaso AOD, Miceli R, Schettino G (2018) In: 2018 AEIT international annual conference, pp 1–6
13. Bergveld HJ, Kruijt WS, Notten PHL (2002) Battery management systems design by modelling. Kluwer Academic Publishers
14. https://www.plm.automation.siemens.com/global/en/products/simcenter/STAR-CCM.html
15. https://www.ansys.com/products/fluids/ansys-fluent
16. Saldana G, San Martin JI, Zamora I, Asensio FJ, Onederra O (2019) Energies 12
17. Raël S, Urbain M, Renaudineau H (2014) In: 2014 IEEE 23rd international symposium on industrial electronics (ISIE), pp 1760–1765
18. Newman J, Tiedemann W (1975) AICHE J 21:25–41
19. Newman JS, Tobias CW (1962) J Electrochem Soc 109:1183
20. Fuller TF, Doyle M, Newman J (1994) J Electrochem Soc 141:982–990
21. Albertus P, Christensen J, Newman J (2009) J Electrochem Soc 156:A606–A618
22. Christensen J, Newman J (2006) J Solid State Electrochem 10:293–319
23. Rao L, Newman J (1997) J Electrochem Soc 144:2697–2704
24. Thomas KE, Newman J (2003) J Power Sources 119:844–849
25. Botte GG, Subramanian VR, White RE (2000) Electrochim Acta 45:2594–2609
26. Santhanagopalan S, Guo Q, Ramadass P, White RE (2006) J Power Sources 156:620–628
27. Verbrugge MW (1995) AICHE J 41:1550–1562
28. Verbrugge MW, Koch BJ (1999) J Electrochem Soc 146:833–839
29. Verbrugge MW, Baker DR, Koch BJ (2002) J Power Sources 110:295–309
30. Verbrugge M, Schlesinger M (2010) Adaptive characterization and modeling of electrochemical energy storage devices for hybrid electric vehicle applications. In: pp 417–524
31. Johnson VH, Pesaran AA, Sack T (2000) In: Proceedings of the 17th Annual Electric Vehicle Symposium, Montrieal, Canada. EVAA, Washington, D.C. OCLC Number: 48393549
32. Pesaran AA (2002) J Power Sources 110:377–382
33. Zolot M, Pesaran AA, Mihalic M (2002) Thermal evaluation of Toyota Prius battery pack. In: Future car congress. SAE International. https://doi.org/10.4271/2002-01-1962. ISSN:0148-7191
34. Stuart T, Fang F, Wang X, Ashtiani C, Pesaran A (2002) SAE Trans 111:777–785
35. Kim G-H, Pesaran A (2007) World Electric Veh J 1:126–133
36. Kim G-H, Pesaran A, Spotnitz R (2007) J Power Sources 170:476–489
37. Kim G-H, Smith K, Lee K-J, Santhanagopalan S, Pesaran A (2011) J Electrochem Soc 158:A955
38. Hutzenlaub T, Thiele S, Paust N, Spotnitz R, Zengerle R, Walchshofer C (2014) Electrochim Acta 115:131–139
39. Marcicki J, Zhu M, Bartlett A, Yang XG, Chen Y, Miller T, L'Eplattenier P, Caldichoury I (2017) J Electrochem Soc 164:A6440–A6448
40. Hanke F, Modrow N, Akkermans RLC, Korotkin I, Mocanu FC, Neufeld VA, Veit M (2019) J Electrochem Soc 167:013522
41. Zhu J, Li W, Wierzbicki T, Xia Y, Harding J (2019) Int J Plast 121:293–311
42. Ping P, Wang Q, Chung Y, Wen J (2017) Appl Energy 205:1327–1344
43. Darcovich K, MacNeil DD, Recoskie S, Cadic Q, Ilinca F (2019) Appl Therm Eng 155:185–195
44. Darcovich K, MacNeil DD, Recoskie S, Kenney B (2018) Appl Therm Eng 133:566–575
45. Logan ER, Dahn JR (2020) Trends Chem 2:354–366
46. Spotnitz R, Gering KL, Hartridge S, Damblanc G (2014) ECS Trans 58:25–36

47. Franco AA, Rucci A, Brandell D, Frayret C, Gaberscek M, Jankowski P, Johansson P (2019) Chem Rev 119:4569–4627
48. Urban A, Seo D-H, Ceder G (2016) npj Comput Mater 2:16002
49. Wang A, Kadam S, Li H, Shi S, Qi Y (2018) npj Comput Mater 4:15
50. Valoen LO, Reimers JN (2005) J Electrochem Soc 152:A882–A891
51. Gering KL (2006) Electrochim Acta 51:3125–3138
52. Gering KL (2017) Electrochim Acta 225:175–189
53. Ramasubramanian A, Yurkiv V, Nie A, Najafi A, Khounsary A, Shahbazian–Yassar R, Mashayek F (2019) Int J Solids Struct 159:163–170
54. Tokur M, Aydin A, Cetinkaya T, Akbulut H (2017) J Electrochem Soc 164:A2238–A2250
55. Clancy T, Rohan JF (2015) J Phys Conf Ser 660:012075
56. Painter R, Sharpe L, Hargrove SK (2017) In: Comsol conference, Boston, MA
57. Spotnitz R, Kaludercic B, Muzaferija S, Peric M, Damblanc G, Hartridge S, Int SAE (2012) J Alt Power 1:160–168
58. Wiedemann AH, Goldin GM, Barnett SA, Zhu H, Kee RJ (2013) Electrochim Acta 88:580–588
59. Zhang D, Bertei A, Tariq F, Brandon N, Cai Q (2019) Prog Energy 1:012003
60. Cooper SJ, Eastwood DS, Gelb J, Damblanc G, Brett DJL, Bradley RS, Withers PJ, Lee PD, Marquis AJ, Brandon NP, Shearing PR (2014) J Power Sources 247:1033–1039
61. Goldin GM, Colclasure AM, Wiedemann AH, Kee RJ (2012) Electrochim Acta 64:118–129
62. ChiuHuang C-K, Huang H-YS (2015) J Solid State Electrochem 19:2245–2253
63. Fan XY, Zhuang QC, Wei GZ, Ke FS, Huang L, Dong QF, Sun SG (2009) Acta Chim Sin 67:1547–1552
64. Wang M (2017) In Mechanical engineering. Michigan State University
65. Harb JN, LaFollette RM (1999) J Electrochem Soc 146:809–818
66. Reimers JN (2013) J Electrochem Soc 161:A118–A127
67. Howard WF, Spotnitz RM (2007) J Power Sources 165:887–891
68. Kwon KH, Shin CB, Kang TH, Kim CS (2006) J Power Sources 163:151–157
69. Madani SS, Swierczynski M, Kær SK (2017) ECS Trans 81:261–270
70. Liaw BY, Nagasubramanian G, Jungst RG, Doughty DH (2004) Solid State Ionics 175:835–839
71. Verbrugge M, Koch B (2006) J Electrochem Soc 153:A187–A201
72. Deng J, Bae C, Miller T, L'Eplattenier P, Bateau-Meyer S (2018) J Electrochem Soc 165:A3067–A3076
73. Yeduvaka G, Spotnitz R, Gering K (2019) ECS Trans 19:1–10
74. Hartridge S, Damblanc G, Spotnitz R, Imaichi K (2011) In: JSAE annual congress. Society of Automotive Engineers of Japan, Yokohama, Japan,
75. Sakti A, Michalek JJ, Chun S-E, Whitacre JF (2013) Int J Energy Res 37:1562–1568
76. Bandla VN (2020) In: Advanced automotive battery conference, virtual
77. Panchal S, Mathew M, Fraser R, Fowler M (2018) Appl Therm Eng 135:123–132
78. Panchal S, Dincer I, Agelin-Chaab M, Fraser R, Fowler M (2017) Int J Heat Mass Transf 109:1239–1251
79. Vyroubal P, Kazda T, Maxa J (2016) ECS Trans 74:3–8
80. Oh K-Y, Epureanu BI (2016) Appl Energy 178:633–646
81. Kleiner J, Komsiyska L, Eiger G, Endisch C (2019) Energies 13:62
82. Li Y, Qi F, Guo H, Guo Z, Li M, Wu W (2019) Case Stud Thermal Eng 13:100387
83. Li G, Li S (2015) ECS Trans 64:1–14
84. Li W, Crompton KR, Hacker C, Ostanek JK (2020) J Energy Storage 32:101890
85. Pan Z, Li W, Xia Y (2020) J Energy Storage 27:101016
86. Deng J, Smith I, Bae C, Rairigh P, Miller T, Surampudi B, L'Eplattenier P, Caldichoury I (2020) J Electrochem Soc 167:090550
87. Sheikh M, Elmarakbi A, Rehman S (2020) J Energy Storage 32:101833
88. Xi S, Zhao Q, Chang L, Huang X, Cai Z (2020) Eng Fail Anal 116:104747
89. Somasundaram K, Birgersson E, Mujumdar AS (2012) J Power Sources 203:84–96

90. Xu M, Zhang Z, Wang X, Jia L, Yang L (2014) J Power Sources 256:233–243
91. Samba A, Omar N, Gualous H, Capron O, Van den Bossche P, Van Mierlo J (2014) Electrochim Acta 147:319–329
92. Falconi A (2017) In: Université Grenoble Alpes, pp 215. https://hal.archives-ouvertes.fr/tel-01676976
93. Frohlich K (2013) In: Technical University of Graz, pp 75. https://diglib.tugraz.at/download.php?id=576a748ace225&location=browse
94. Cai L, White RE (2011) J Power Sources 196:5985–5989
95. Hosseinzadeh E, Marco J, Jennings P (2017) Energies 10
96. Dai M, Huo C, Zhang Q, Khan K, Zhang X, Shen C (2018) Adv Theory Simul 1:1800023
97. Gao Y, Zhang X, Yang J, Guo B, Zhou X (2019) Int J Electrochem Sci 14:3180–3203
98. Bizeray AM, Zhao S, Duncan SR, Howey DA (2015) J Power Sources 296:400–412
99. Farrag ME, Haggag A, Parambu RB, Zhou C (2017) In: 2017 Nineteenth international middle east power systems conference (MEPCON), pp 677–682
100. Singh N, Khare N, Chaturvedi PK (2015) In: COMSOL conference, Pune
101. Singh P, Khare N, Chaturvedi PK (2018) Int J Eng Sci Technol 21:35–42
102. Painter R, Berryhill B, Sharpe L, Hargrove SK (2014) In: COMSOL conference, Boston, MA
103. Rajabloo B, Desilets M, Choquette Y (2015), In: Comsol 2015, Boston, MA
104. Lam LL, Darling RB (2011) In: Comsol 2011, Boston, MA
105. Tang S, Liu Y, Li L, Jia M, Jiang L, Liu F, Ai Y, Yao C, Gu H (2020) Ionics
106. Li C, Zhang H, Zhang R, Lin Y, Fang H (2021) Appl Therm Eng 182:116144
107. Chiew J, Chin CS, Toh WD, Gao Z, Jia J, Zhang CZ (2019) Appl Therm Eng 147:450–463
108. Tan M, Gan Y, Liang J, He L, Li Y, Song S, Shi Y (2020) Appl Therm Eng 177:115500
109. Li H, Liu B, Zhou D, Zhang C (2020) J Electrochem Soc 167:120501
110. Sripad S, Viswanathan V (2017) J Electrochem Soc 164:E3635–E3646
111. Dareini A (2016) In: Blekinge Institute of Technology, Karlskrona, Sweden, pp 119
112. Tanim TR (2015) In: Mechanical engineering. The Pennsylvania State University, pp 137
113. Wang Q, Shaffer CE, Sinha PK (2015) Front Energy Res 3:35
114. Yang X-G, Leng Y, Zhang G, Ge S, Wang C-Y (2017) J Power Sources 360:28–40
115. Yin H, Ma S, Li H, Wen G, Santhanagopalan S, Zhang C (2021) eTransportation 7:100098
116. Li W, Zhu J (2020) J Electrochem Soc 167:120504
117. Zhu J, Luo H, Li W, Gao T, Xia Y, Wierzbicki T (2019) Int J Impact Eng 131:78–84
118. Lian J, Wierzbicki T, Zhu J, Li W (2019) Eng Fract Mech 217:106520
119. Schmalstieg J, Rahe C, Ecker M, Sauer DU (2018) J Electrochem Soc 165:A3799–A3810
120. Ecker M, Käbitz S, Laresgoiti I, Sauer DU (2015) J Electrochem Soc 162:A1849–A1857
121. Santhanagopalan S, Smith K, Neubauer J, Kim G-H, Keyser M, Pesaran AA, Artech H (2015) Design and analysis of large lithium-ion battery systems. Artech House, Boston/London
122. Deng J, Bae C, Marcicki J, Masias A, Miller T (2018) Nat Energy 3:261–266
123. Kawai T (2020) In: Advanced automotive battery conference Europe, Wiesbaden, Germany
124. Basu S, Hariharan KS, Kolake SM, Song T, Sohn DK, Yeo T (2016) Appl Energy 181:1–13
125. Deng T, Zhang G, Ran Y, Liu P (2019) Appl Therm Eng 160:114088
126. Deng T, Ran Y, Zhang G, Yin Y (2019) Appl Therm Eng 150:1186–1196
127. Deng T, Ran Y, Zhang G, Chen X, Tong Y (2019) Int J Heat Mass Transf 139:963–973
128. Sheng L, Su L, Zhang H, Li K, Fang Y, Ye W, Fang Y (2019) Int J Heat Mass Transf 141:658–668
129. Lajunen A, Kalttonen A (2015) In: 2015 IEEE transportation electrification conference and expo (ITEC), pp 1–6
130. Tran T-H, Harmand S, Sahut B (2014) J Power Sources 265:262–272
131. Tran T-H, Harmand S, Desmet B, Filangi S (2014) Appl Therm Eng 63:551–558
132. Wang J, Pan C, Xu H, Xu X (2017) In: 5th international conference on mechanical, automotive and materials engineering (CMAME), pp 223–228
133. Wang J, Xu H, Xu X, Pan C (2017) IOP Conf Ser: Mater Sci Eng 231:012025
134. Vora AP (2016) In: Purdue University, pp 197. https://docs.lib.purdue.edu/open_access_dissertations/875

135. Berglund E (2020) In: Mechanical engineering. Chalmers University of Technology, pp 36, https://odr.chalmers.se/bitstream/20.500.12380/300718/1/2020-02%20Emma%20Berglund.pdf

136. Cheng M, Feng L, Chen B (2015) In: SAE International, Paper No. 2015-01-1198

137. Xing J, Vora AP, Hoshing V, Saha T, Shaver GM, Wasynczuk O, Varigonda S (2017) In Proceedings of the 2017 American Control Conference (ACC), pp 74–79

138. Carnovale A (2016) In: Mechanical and mechatronics engineering. University of Waterloo, pp 110. https://core.ac.uk/download/pdf/144149891.pdf

139. Li X, He F, Ma L (2013) J Power Sources 238:395–402

140. Liu Y, Liao YG, Lai M-C (2019) Vehicles 1:127–137

141. Sun H, Wang X, Tossan B, Dixon R (2012) J Power Sources 206:349–356

142. Xun J, Liu R, Jiao K (2013) J Power Sources 233:47–61

143. He F, Li X, Ma L (2014) Int J Heat Mass Transf 72:622–629

144. Panchal S, Khasow R, Dincer I, Agelin-Chaab M, Fraser R, Fowler M (2017) Numer Heat Transf Part A: Appl 71:626–637

145. Wang S, Ji S, Zhu Y (2021) Appl Therm Eng 182:116040

146. Tang A, Li J, Lou L, Shan C, Yuan X (2019) Appl Therm Eng 159:113760

147. Zhao C, Cao W, Dong T, Jiang F (2018) Int J Heat Mass Transf 120:751–762

148. Zhao C, Sousa ACM, Jiang F (2019) Int J Heat Mass Transf 129:660–670

149. Xie L, Huang Y, Lai H (2020) Appl Therm Eng 178:115599

150. Li Y, Zhou Z, Wu W-T (2019) Appl Therm Eng 147:829–840

151. Li W, Xiao M, Peng X, Garg A, Gao L (2019) Appl Therm Eng 147:90–100

152. Zhou H, Zhou F, Xu L, Kong J, Qingxin Yang (2019) Int J Heat Mass Transf 131:984–998

153. Wang T, Tseng KJ, Zhao J, Wei Z (2014) Appl Energy 134:229–238

154. Chen S, Peng X, Bao N, Garg A (2019) Appl Therm Eng 156:324–339

155. Chen K, Wu W, Yuan F, Chen L, Wang S (2019) Energy 167:781–790

156. Yang T, Yang N, Zhang X, Li G (2016) Int J Therm Sci 108:132–144

157. Na X, Kang H, Wang T, Wang Y (2018) Appl Therm Eng 143:257–262

158. Wang Y, Zhang G, Yang X (2019) Appl Therm Eng 162:114200

159. Jiaqiang E, Han D, Qiu A, Zhu H, Deng Y, Chen J, Zhao X, Zuo W, Wang H, Chen J, Peng Q (2018) Appl Therm Eng 132:508–520

160. Song L, Zhang H, Yang C (2019) Int J Heat Mass Transf 133:827–841

161. Liu Y, Zhang J (2019) Appl Energy 252:113426

162. Cao W, Zhao C, Wang Y, Dong T, Jiang F (2019) Int J Heat Mass Transf 138:1178–1187

163. Yang X-H, Tan S-C, Liu J (2016) Energy Convers Manag 117:577–585

164. Chung Y, Kim MS (2019) Energy Convers Manag 196:105–116

165. Lai Y, Wu W, Chen K, Wang S, Xin C (2019) Int J Heat Mass Transf 144:118581

166. Zhang H, Li C, Zhang R, Lin Y, Fang H (2020) Appl Therm Eng 173:115216

167. Zhou H, Zhou F, Zhang Q, Wang Q, Song Z (2019) Appl Therm Eng 162:114257

168. Kizilel R, Sabbah R, Selman JR, Al-Hallaj S (2009) J Power Sources 194:1105–1112

169. Li X, He F, Zhang G, Huang Q, Zhou D (2019) Appl Therm Eng 146:866–880

170. Yang N, Zhang X, Li G, Hua D (2015) Appl Therm Eng 80:55–65

171. Mohammed AH, Esmaeeli R, Aliniagerdroudbari H, Alhadri M, Hashemi SR, Nadkarni G, Farhad S (2019) Appl Therm Eng 160:114106

172. Al-Zareer M, Dincer I, Rosen MA (2019) Energy Convers Manag 187:191–204

173. Bugryniec PJ, Davidson JN, Brown SF (2020) Energy Rep 6:189–197

174. Barsukov Y, Qian J (2013) Battery power management for portable devices. Artech House, Norwood

175. Hannan MA, Lipu MSH, Hussain A, Mohamed A (2017) Renew Sust Energ Rev 78:834–854

176. Ali MU, Zafar A, Nengroo SH, Hussain S, Junaid Alvi M, Kim H-J (2019) Energies 13:446

177. Liu X, Ma Y, Ying Z (2013) In: Proceedings of the 32nd Chinese control conference, pp 7680–7685

178. Chen Y, Ma Y, Chen H (2018) J Renew Sustain Energy 10:034103

179. Locorotondo E, Pugi L, Berzi L, Pierini M, Pretto A (2018) In: 2018 IEEE international conference on environment and electrical engineering and 2018 IEEE industrial and commercial power systems Europe (EEEIC/I&CPS Europe), pp 1–6
180. Veeraraghavan A, Adithya V, Bhave A, Akella S (2017) In: 2017 IEEE transportation electrification conference (ITEC-India), pp 1–4
181. Lee U, Kang N, Lee I (2019) Struct Multidiscip Optim 60:949–963
182. Mahmud K, Town GE (2016) Appl Energy 172:337–359
183. Ali ZK, Badjate SL, Kshirsagar RV (2015) Int J Adv Trends Comput Sci Eng 5:1–5
184. Ringkjøb H-K, Haugan PM, Solbrekke IM (2018) Renew Sust Energ Rev 96:440–459
185. Wang Y, Gao Q, Zhang T, Wang G, Jiang Z, Li Y (2017) Energies 10:1636
186. Park G, Lee S, Jin S, Kwak S (2014) Expert Syst Appl 41:2595–2607
187. Liu J, Dong H, Jin T, Lu L, Manouchehrinia B, Dong Z (2018) Energies 11:2699
188. McWhirter TE, Wagner TJ, Stubbs JE, Rizzo DM, Williams JB (2020) Int J Electric Hybrid Veh 12:1
189. Broglia L, Autefage B, Ponchant M (2012) World Electric Veh J 5:1082–1089
190. Badin F, Le Berr F, Briki H, Dabadie J-C, Petit M, Magand S, Condemine E (2013) World Electric Veh J 6:112–123
191. Irimia C, Grovu M, Sirbu G, Birtas A, Husar C, Ponchant M (2019) In: 2019 electric vehicles international conference (EV), pp 1–6
192. Vatanparvar K, Wan J, Faruque MAA (2015) In: 2015 IEEE/ACM international symposium on low power electronics and design (ISLPED), pp 353–358
193. Faruque MAA, Vatanparvar K (2016) In: 2016 21st Asia and South Pacific design automation conference (ASP-DAC), pp 423–428
194. Zhang J, Yuan Y, Lv C, Li Y (2015) In: 2015 IEEE international conference on mechatronics and automation (ICMA), pp 1307–1312
195. Daowei Z, Hongrui C, Shishun Z, Gang Y (2014) In: 2014 IEEE conference and expo transportation electrification Asia-Pacific (ITEC Asia-Pacific), pp 1–4
196. Husar C, Grovu M, Irimia C, Desreveaux A, Bouscayrol A, Ponchant M, Magnin P (2019) In: 2019 IEEE vehicle power and propulsion conference (VPPC), pp 1–6
197. Berzi L, Favilli T, Pierini M, Pugi L, Weiß GB, Tobia N, Ponchant M (2019) In: 2019 IEEE 5th international forum on research and technology for society and industry (RTSI), pp 308–313
198. Li A, Ponchant M, Sturm J, Jossen A (2020) World Electric Veh J 11:75
199. Caldichoury I, Leplattenier P (2021) In: AIAA scitech 2021 forum. American Institute of Aeronautics and Astronautics. https://doi.org/10.2514/6.2021-1278
200. Wu B, Lu W (2016) J Electrochem Soc 163:A3131–A3139

Chapter 2
In Situ Measurement of Current Distribution in Large-Format Li-Ion Cells

Guangsheng Zhang, Christian E. Shaffer, Xiao Guang Yang, Christopher D. Rahn, and Chao-Yang Wang

Abstract Non-uniform current distributions in large-format Li-ion cells can cause underutilization of active materials, reduction of usable energy density, non-uniform heat generation, exacerbated lithium plating, and accelerated degradation. In situ measurements of current distributions in large-format Li-ion cells not only reveal local behaviors but also provide spatially resolved data for validation of electrochemical-thermal coupled models. The insights and models aid the development of faster rechargeable, energy denser, safer, and more durable Li-ion batteries. This work reviews the progress of in situ measurement of current distribution over the electrode area for large-format Li-ion cells. Direct measurement using segmented cells, indirect measurement using embedded local potential tabs, and noninvasive diagnosis with magnetic resonance imaging are discussed. Key findings from the measurements are then summarized, such as the current distributions under various operating conditions and cell designs, and the effects of non-uniform current

---

Statement Parts of this chapter are reproduced with permission from Zhang et al. (2013) *J. Electrochem. Soc.* 160, A610 and Zhang et al. 2013 *J. Electrochem. Soc.* 160, A2299, Copyright 2013 The Electrochemical Society.

G. Zhang (✉)
Department of Mechanical & Aerospace Engineering, The University of Alabama in Huntsville, Huntsville, AL, USA
e-mail: gz0002@uah.edu

C. E. Shaffer · X. G. Yang
Electrified Powertrain Engineering, Ford Motor Company, Dearborn, MI, USA

C. D. Rahn
Battery and Energy Storage Technology Center and Department of Mechanical Engineering, The Pennsylvania State University, University Park, PA, USA

C.-Y. Wang
Electrochemical Engine Center (ECEC) and Department of Mechanical Engineering, The Pennsylvania State University, University Park, PA, USA

S. Santhanagopalan (ed.), *Computer Aided Engineering of Batteries*,
Modern Aspects of Electrochemistry 62,
https://doi.org/10.1007/978-3-031-17607-4_2

distributions on local state of charge and usable energy density. Finally, future research needs are proposed.

2.1 Introduction

Since its commercialization in the early 1990s, Li-ion battery technology quickly dominated and advanced the portable electronics industry with advantages of high energy density, low self-discharge, and long cycle life [1–2]. In the past decade, its application in electric vehicles has proliferated to enable a sustainable energy future [3]. As compared to applications in portable electronics, Li-ion batteries in electric vehicles are much larger in size and more demanding in terms of energy density, power density, safety, durability, and cost. Although much progress has been made, grand challenges remain [4–5]. In particular, how to unlock the potential of Li battery materials and to scale up Li-ion cells to hundreds of Ah per cell without substantial losses in power performance, durability, and safety remains a key technological challenge. This challenge is exacerbated by the increasing demand of fast charging [6–9], low temperature operation [10], high energy density, and robust safety [11–13] driven by wider adoption of electric vehicles in recent years.

As schematically shown in Fig. 2.1, in large-format Li-ion cells, especially cylindrical and prismatic ones with long current collectors and limited tabs [14–15], the path of current through thin metal foil current collectors is not equal at different locations along the electrodes. Through complex interactions among the local reaction current, state of charge (SOC), and temperature, non-uniform spatial distribution of these parameters is inevitable, especially during high rate charging and discharging. Such non-uniform distributions will lead to underutilization of active materials loaded in a large cell, thereby drastically reducing its energy density from the coin cell benchmark [15]. Of equal importance, non-uniform current distribution can exacerbate localized lithium plating during fast charging and localized overheating during high rate operation, compromising durability and safety of large batteries. Therefore, understanding how these critical parameters distribute under various design and operation conditions greatly aids development of faster rechargeable, energy denser, safer, and more durable Li-ion batteries.

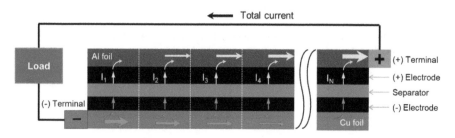

Fig. 2.1 Schematic of current flow path in a Li-ion cell with a pair of counter located terminals during discharging (the schematic is not to scale)

Modeling has been widely used in the research and development of Li-ion cells to gain insight into electrochemical-thermal coupled phenomena, predict cell performance and aging, optimize cell design and operation, and accelerate development of new concepts and strategies [15–26]. But these models need to be validated against experimental data to be reliable. Considering that critical internal parameters of local current density, SOC, and temperature are non-uniformly distributed spatially and temporally in large-format Li-ion cells, multidimensional models need to be validated against not only overall performance data but also spatially distributed data.

Therefore, it is important to experimentally measure current density, SOC, and temperature in large-format Li-ion cells locally, in order to reveal insight about their distribution characteristics and to provide spatially resolved data for multi-dimensional model validation. Considering that distribution of those parameters is transient in nature, in situ measurement is desired.

Supported by US Department of Energy's CAEBAT program, a research team at EC Power and The Pennsylvania State University (PSU) measured current distributions in situ along electrode length using a segmented cell approach [27–28]. Based on the measured current distributions, SOC distributions were obtained. The team also in situ measured temperature distributions in Li-ion cells [29–30]. The measurement revealed many insights on Li-ion cells and provided spatially resolved data for model validation [15]. Since then other work on current distribution measurement have been reported using Li-ion cells with embedded local potential measurement tabs [31–33] or using noninvasive magnetic resonance imaging (MRI) for commercial Li-ion cells [34]. However, further efforts are still needed on this important topic. In particular, behaviors of current distribution under extreme conditions with emerging battery materials or designs are not well understood in spite of their importance to wide adoption of electric vehicles, such as extreme fast charging [9], low temperature operation [10], and safety abuse conditions [35].

Note that non-uniformity of current distribution in Li-ion batteries exists at various levels, including module level consisting of parallel-connected cells [36–37], cell level along electrode length [27–28, 33–34, 38], and electrode level across electrode thickness [39–41]. In this work, we focus on in situ measurement of current distribution at cell level along electrode length which is special to large-format Li-ion cells. In Sect. 2.2, we review the work of direct measurement using segmented Li-ion cells by the EC Power-PSU team. In Sect. 2.3, we review the work of indirect measurement using embedded local potential tabs [31–33]. In Sect. 2.4, we review the work of noninvasive diagnosis with magnetic resonance imaging [34]. In Sect. 2.5, we briefly summarize the progress and propose further efforts to address the emerging needs of electric vehicle Li-ion cells.

2.2 Direct Measurement of Current Distribution Using Segmented Li-Ion Cells

As indicated in Fig. 2.1, if one or both electrodes in a Li-ion cell are electrically segmented, for example, along the dashed lines, local currents in each segment would be forced to flow separately, allowing direct measurement of current distribution. The EC Power-PSU team took this approach. This section is a summary of the work by the team, including both published [27–28] results and previously unpublished results (Sects. 2.2.2.6, 2.2.2.7 and 2.2.2.8).

2.2.1 Experimental Method Using Segmented Li-Ion Cells

2.2.1.1 Experimental Cell with Segmented Positive Electrode

Figure 2.2 shows schematically the experimental cell used in studies by the EC Power-PSU team [27–28]. The experimental cell consists of one intact negative electrode (with active material coated on both sides of a Cu foil), two layers of separators, and ten positive electrode segments (with active material coated on both sides of an Al foil). The negative electrode has only one tab, which serves as the negative terminal of the cell. Every positive electrode segment has two tabs, with only one tab per segment used in study [27] and both tabs per segment used in study [28]. The intact negative electrode and separators are folded in a serpentine manner, (Z-fold) with one positive electrode segment sandwiched inside each fold.

In the experimental cell, positive and negative electrode active materials are lithium iron phosphate (LFP) and graphite, respectively. The separator is Celgard® 2320 PP/PE/PP membrane (20 μm thick). The electrolyte is 1.2 M LiPF$_6$ in EC:EMC:PC (45:50:5 v%). The positive electrode coating (each side) is 64 μm thick, and negative electrode is 43 μm thick. The Al foil current collector is 15 μm thick, and the Cu foil is 10 μm thick. Every positive electrode segment is 150 mm long and 56.5 mm wide (the coated area). The negative electrode is ~1.8 m long. The electrode-separator sandwich is assembled in a pouch. After electrolyte filling, the pouch cell is sealed with 1 negative terminal and 20 positive terminals exposed to the outside. The cell has a nominal capacity of 2.4 Ah.

2.2.1.2 Experimental System

Figure 2.3 shows a schematic of the experimental system in study [27]. Note that the experimental Li-ion cell is shown unfolded and simplified in the schematic for clarity. The ten positive electrode segments are connected in parallel to a low-resistance bus wire, each through a shunt resistor (PLV7AL, 2 mΩ ± 0.5%, Precision Resistor Co., USA) for local current sensing. A low resistance meter (3560, Hioki, Japan) is used to ensure that the resistance of shunt and connecting

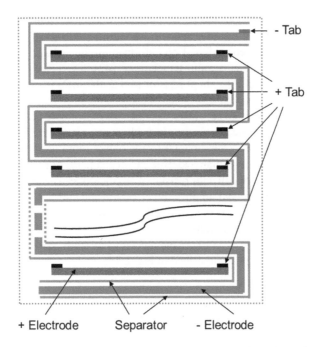

Fig. 2.2 Schematic of experimental Li-ion cell with segmented positive electrode (not to scale). (Adapted with permission from Ref. [27] Copyright 2013 The Electrochemical Society)

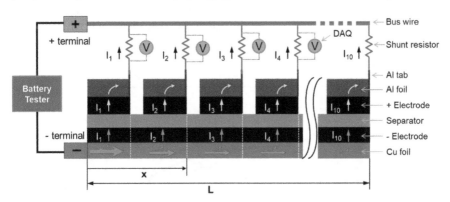

Fig. 2.3 Schematic of experimental system for current distribution measurement (The experimental Li-ion cell is shown unfolded and simplified in the schematic for clarity; arrows represent current flow during discharge). (Reused with permission from Ref. [27] Copyright 2013 The Electrochemical Society)

wires is the same for every channel (4.0 ± 0.1 mΩ) and resistance of the bus wire is negligible (<0.1 mΩ). By connecting the positive electrode segments in parallel and making resistance from the positive terminal bus to the Al tab uniform for each local channel, the effects of Al foil on current distribution can be suppressed so as to focus on the effects of the Cu foil current collector. Note that DC internal resistance of each cathode segment and corresponding anode is estimated to be more than 800 mΩ using a method based on cell open circuit voltage and cell voltage at 10 s into C/5 discharge at room temperature [42]. Therefore, the influence of additional resistance from the shunt resistor (4.0 ± 0.1 mΩ) is negligibly small as compared with the DC internal resistance of each segment. Also note that in present-day designs, commercial Li-ion cells with negative electrodes as long as 1.8 m would rarely have only one tab for the electrode due to excessive resistance. So the current distribution in this experimental Li-ion cell is more non-uniform than that in a commercial cell with similar size. But the mechanisms of non-uniform current distribution due to current collector resistance are the same. The experimental setup in this study can separate and amplify the effects of the current collector to reveal insights on current distribution in Li-ion cells. Even for commercial Li-ion cells, the resistance of current collectors can influence current distribution during high current charging and discharging. The experimental setup in study [28] has different tab configurations, which will be described in Fig. 2.17.

2.2.2   Results from Segmented Li-Ion Cell

2.2.2.1   Overall Cell Performance

Figure 2.4 shows voltage of the experimental cell during discharge at various C rates [27]. Open circuit voltage (OCV) of the cell at different stages of discharge is also shown, which are obtained by discharging the cell at C/10 intermittently followed

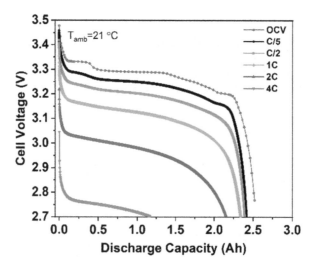

Fig. 2.4 Voltage of the experimental Li-ion cell during discharge at various C rates at room temperature (21 °C). (Adapted with permission from Ref. [27] Copyright 2013 The Electrochemical Society)

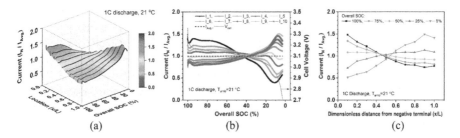

Fig. 2.5 Spatial-temporal current distribution during 1C discharge at 21 °C. (**a**) Variation of local currents with both location and overall SOC; (**b**) Variation of local currents and cell voltage with overall SOC; (**c**) Spatial current distribution at different SOC. (Adapted with permission from Ref. [27] Copyright 2013 The Electrochemical Society)

by a rest period of 1 h. The cell voltage and discharge capacity decreased with increasing C rate, which was attributed to higher ohmic and kinetic overvoltage at higher current. The OCV variation was attributed primarily to open circuit potential (OCP vs. Li/Li$^+$) variation of graphite [43] because LFP has a flat OCP curve in a wide range of state of charge (SOC) [44].

2.2.2.2 Current Distribution During 1 C Discharge at Room Temperature

Figure 2.5a shows the measured spatial-temporal current distribution during 1C discharge at room temperature (21 °C). Note that local currents are made dimensionless after being normalized by the average current for convenient comparison among different cases. The discharge time is made dimensionless by using the cell overall SOC to indicate discharge progress. Location of each segment is made dimensionless using relative distance from the negative terminal of the experimental Li-ion cell. Figure 2.5b shows temporal variation of each local current, with average current (equal to the unity after normalization) and cell voltage also shown in the figure. Figure 2.5c shows spatial distribution of local currents at different levels of cell SOC, using the local currents data extracted from Fig. 2.5b. Current distribution is not uniform from beginning of discharge and evolves significantly as discharge proceeds. Initially, segments closer to the negative terminal produce higher local currents than those farther away, i.e., $I_1 > I_2 > I_3 \ldots > I_{10}$. As discharge continues, local currents in segments with higher initial values generally decrease, while those with lower initial values increase, which leads to an entirely different pattern of current distribution near the end of discharge ($I_1 < I_2 < I_3 \ldots$). Note that I_9 is almost always slightly higher than I_{10}, which was attributed to that segment 9 had slightly higher capacity than other segments [27].

The very non-uniform current distribution at the beginning of discharge was attributed to the resistance of the negative current collector (~55 mΩ) [27]. As schematically shown in Fig. 2.3, current flows in the Cu foil from the negative terminal to the other end and gradually decreases along the flow direction as local currents enter the negative electrode. As the current flows in, considerable potential drop along the Cu foil occurs due to its resistance. On the other hand, potentials of segments on the positive side (joint of shunt resistor with bus wire) are essentially equal due to negligible resistance of the bus wire. Consequently, the local overvoltage is smaller for segments farther from the negative terminal, thereby driving lower current according to electrochemical kinetics (e.g., Butler-Volmer equation) [45]. Therefore, the farther a segment is from the negative terminal, the smaller the local overvoltage is, and the lower current it generates. Higher SOC remains in regions far away from the negative terminal.

As discharge proceeds, the regions farther away from the negative terminal maintain higher SOC, and thus higher OCV, due to less current generation earlier and hence produce higher local currents later. The dramatic variation of local currents during late stages of discharge can be attributed to the counteracting effects between potential drop along the Cu foil and the local SOC (or OCV) non-uniformity. The two effects counteract, tending to balance the current distribution. Therefore, local currents in segments with higher initial values would decrease, while local currents in segments with lower initial values would increase. When the effect of local SOC (or OCV) overrides the effect of the ohmic potential drop along the Cu foil, in the case near the end of discharge, the current distribution would reverse in pattern, becoming entirely different from that at the beginning of discharge. As will be shown in Fig. 2.11a, local SOC during discharge can be estimated from local currents data, and the results agree with this explanation very well.

2.2.2.3 Effects of Discharging C Rate on Current Distribution

Discharge C rate was found to have significant effects on current distribution [27]. Figures 2.6 and 2.7 show results of current distribution during C/5 and 4C

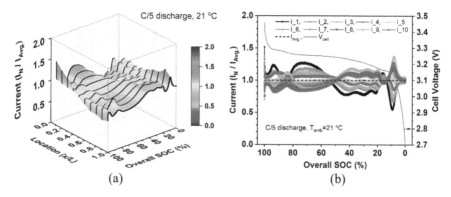

Fig. 2.6 Current distribution during C/5 discharge at 21 °C. (**a**) Variation of local currents with both location and overall SOC; (**b**) Variation of local currents and cell voltage with overall SOC. (Adapted with permission from Ref. [27] Copyright 2013 The Electrochemical Society)

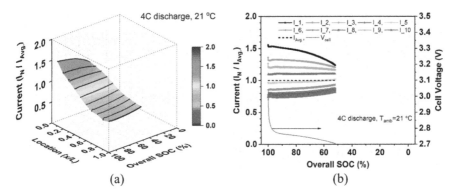

Fig. 2.7 Current distribution during 4C discharge at 21 °C. (**a**) Variation of local currents with both location and overall SOC; (**b**) Variation of local currents and cell voltage with overall SOC. (Adapted with permission from Ref. [27] Copyright 2013 The Electrochemical Society)

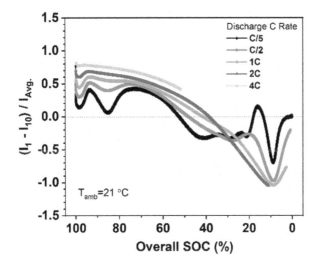

Fig. 2.8 Maximum difference in local current during different C rate discharge at 21 °C. (Adapted with permission from Ref. [27] Copyright 2013 The Electrochemical Society)

discharge, respectively, for comparison with results at 1C. It can be seen clearly that local currents spread less during lower C-rate discharge, indicating more uniform current distribution. This trend can be more clearly seen from Fig. 2.8, in which the differences between I_1 and I_{10} during different C rate discharge are plotted together. Segments 1 and 10 are most apart along the negative current collector, and I_1 and I_{10} are most different, so their difference, i.e., the maximum ΔI, can be used to represent the non-uniformity of current distribution in the experimental cell.

The effects of C-rate on current distribution were also attributed to the opposing effects of potential drop along the Cu foil and the local SOC (or OCV) non-uniformity [27]. During low C rate discharge (C/5), potential drop along the Cu foil is so small that the local current distribution is controlled by local SOC (or OCV) non-uniformity. Indeed, a wavy pattern, evident from Figs. 2.6 and 2.8 results, reflective of multiple plateaus and frequent slope changes seen in the OCV vs. SOC curve. During high C rate discharge (4C), the potential drop along the Cu foil, amounting to ~20 times larger than that during C/5 discharge, would dominate the local current distribution. This dominance by the ohmic potential drop creates smoothly varying patterns, as can be seen clearly from Fig. 2.8 for the cases of 1C, 2C, and 4C. It was hypothesized that a battery operated in close vicinity to the equilibrium features a wavy current distribution, while a battery operated far away from the equilibrium would exhibit a monotonic variation.

2.2.2.4 Effects of Ambient Temperature

Ambient temperature was also found to have significant influences on current distributions [27]. Figures 2.9 and 2.10 show the current distribution at 0 °C and 45 °C during 1C discharge. Comparison with the results at 21 °C, as shown in Fig. 2.5, clearly indicates lower overall performance but more uniform current distribution at lower ambient temperature. This interesting phenomenon can be mainly attributed to the effects of temperature on the internal resistance of Li-ion battery cells, which includes the ohmic resistance, the charge transfer resistance, the resistance of SEI, and mass transport resistances. Previous studies show that the charge transfer resistance, SEI resistance [46–47], and electrolyte resistance [48] increase dramatically at lower temperature while the ohmic resistance of the Al and Cu foils change little (actually decrease slightly [49]). Using open circuit voltage and cell voltage at 10 s of discharge, it can be estimated that internal resistance of the cell at 0 °C is approximately 60% higher than that at 21 °C. As suggested in Fig. 2.3, higher battery sandwich resistance at lower temperature would make the effects of ohmic potential drop along the Cu and Al foils less significant on the

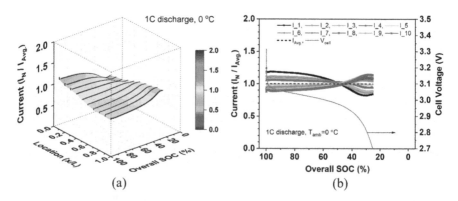

(a)   (b)

Fig. 2.9 Current distribution during 1C discharge at 0 °C. (**a**) Variation of local currents with both location and overall SOC; (**b**) Variation of local currents and cell voltage with overall SOC. (Adapted with permission from Ref. [27] Copyright 2013 The Electrochemical Society)

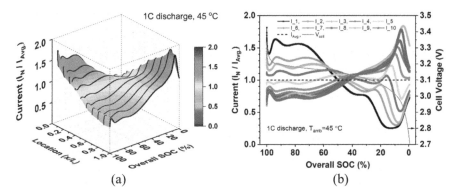

Fig. 2.10 Current distribution during 1C discharge at 45 °C. (**a**) Variation of local currents with both location and overall SOC; (**b**) Variation of local currents and cell voltage with overall SOC. (Adapted with permission from Ref. [27] Copyright 2013 The Electrochemical Society)

current distribution. Therefore, the current distribution would be more uniform at lower temperature. Obviously, higher internal resistance would reduce the overall performance of the experimental cell due to higher overvoltage at the same current. The cell resistance at 45 °C is estimated to be approximately 25% lower than that at 21 °C, so the better overall performance yet more non-uniform current distribution at 45 °C also agrees with the explanations.

2.2.2.5 Local SOC Distribution Calculated from Current Distribution Data

With information on local currents, discharged capacity of every segment was calculated by integrating local current in time from beginning of discharge. Then local SOC was obtained by taking capacity of local segment during C/5 discharge as reference of SOC = 100% [27]. Figure 2.11a shows distribution of local SOC at different overall SOC during 1C discharge. As expected, the SOC distribution first becomes more and more non-uniform as discharge proceeds due to the higher local currents closer to the negative terminal, and then the SOC distribution becomes less non-uniform as discharge approaches end of discharge due to the reversal of current distribution.

It is interesting to note that the SOC distribution is still not uniform at cutoff for 1C discharge. This phenomenon is even more obvious at higher C rates, as can be seen from Fig. 2.11b, which shows SOC distribution at cutoff for different C rate discharge. For the 4C discharge, local SOC of segment 1 is around 30%, while that of segment 10 is higher than 60%. Such highly non-uniform SOC distribution at cutoff significantly underutilizes the active materials loaded in the battery, thereby reducing its energy density. Indeed, the very high SOC of segment 10 suggests that active materials in certain locations are largely underutilized, a significant waste of battery materials.

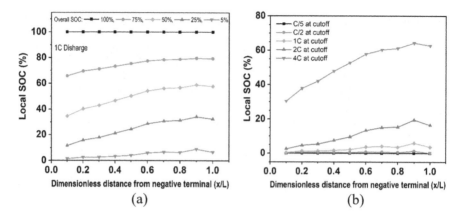

Fig. 2.11 (**a**) Local SOC distribution at different overall SOC during 1C discharge at 21 °C; (**b**) Local SOC distribution at cutoff during different C rate discharge at 21 °C. (Adapted with permission from Ref. [27] Copyright 2013 The Electrochemical Society)

2.2.2.6 Internal Balancing Current After Discharge and Its Effects on Local SOC Distribution

Li-ion cell OCV depends on SOC. Figure 2.12 shows such dependence for the experimental Li-ion cell. It is clear that the OCV generally decreases with SOC except a few plateaus as typical for LFP-Graphite cell. The non-uniform local SOC distribution at cutoff shown in Fig. 2.11 suggests non-uniform local OCV. As schematically shown in Fig. 2.12, such local OCV difference would cause internal balancing current flow. For example, the local SOC of segment 1 is around 30%, while that of segment 10 is higher than 60% at the end of 4C discharge. The local OCV of segment 1 would be nearly 40 mV lower than that of segment 10. Such an OCV difference would drive internal current flow between segment 1 and segment 10 until their OCV become the same. During the balancing process, segment 1 would be charged, while segment 10 is being discharged. Similar phenomena would occur to other segments depending on the OCV difference.

The internal balancing currents were indeed observed in the experiments. Figure 2.13a, b shows the internal balancing current after 1C discharge and 4C discharge, respectively. To reflect the magnitude of internal balancing current, local C rate is used. Note that during balancing, the total current is zero, while local currents are negative or positive. Positive local current means the corresponding segment is being discharged, while a negative local current means the segment is being charged. It can be seen that in both cases, segments closer to negative terminal (1–4) are charged, while segments further away from the negative terminal (6–10) are discharged, corresponding to the non-uniform SOC distribution shown in Fig. 2.11. It is interesting to note that internal balancing takes longer time after higher C rate discharge. After 1C discharge, it took less than 1200 s for all local internal balancing currents to become smaller than 1 mA (C/240). In comparison, after 4C discharge,

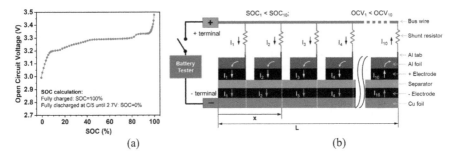

Fig. 2.12 (**a**) OCV of the experimental Li-ion cell at different SOC; (**b**) Schematic of internal balancing current flow in experimental Li-ion cell after discharge

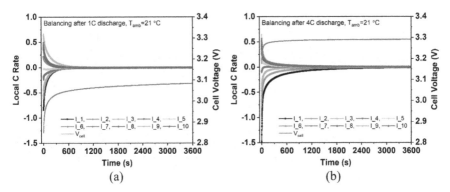

Fig. 2.13 Local internal balancing currents and cell voltage (**a**) after 1C discharge and (**b**) after 4C discharge

the internal balancing current in segment 1 was still larger than 1 mA after 3600 s of rest. It is also interesting to note that the recovery of OCV corresponds to decrease of internal balancing current, suggesting that internal balancing current plays an important role in OCV recovery during rest.

As expected, the internal balancing currents reduce non-uniformity of local SOC distribution. Figure 2.14 shows the cell local SOCs at cutoff voltage (solid lines) and after balancing (dashed lines, after all local currents become smaller than 1 mA). Notably, in Fig. 2.14, the 1C green curves correspond with the resting highlighted in Fig. 2.13a, and the 4C magenta curves correspond with the resting shown in Fig. 2.13b. For the 1C discharge, we observe in Fig. 2.14 that the cell SOC is nearly uniform after balancing. However, it is important to note in Fig. 2.13a that after 3600 s of rest, the cell voltage is still recovering toward ~3.1 V at a rate of 5 μV/s. The relatively quick balancing of the local SOC after 1C discharge (compared with the 4C discharge) can no doubt be attributed in part to the fact that the cell was not significantly out of balance at the end of the 1C discharge. However, the observation that the voltage is still recovering as the cell nears a state of balance is an interesting one. This can most likely be attributed to the significant slope

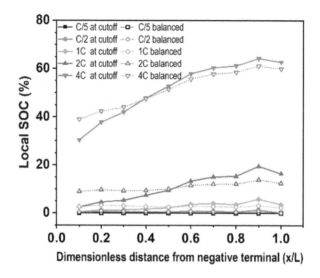

Fig. 2.14 Local SOC distribution in the experimental Li-ion cell before and after balancing

of the OCV-SOC curve at ~3.1 V (see Fig. 2.12), i.e., a small change in SOC is accompanied by a large change in OCV. Compare this observation with the resting after 4C discharge. As shown in Fig. 2.13b and magenta curves in Fig. 2.14, the cell remains at a significantly imbalanced after 3600 s of rest for 4C discharge, even though the voltage recovery rate (dV/dt) is as small as 0.5 μV/s at 3600 s. This behavior after 4C discharge can most likely be attributed to the shape of the OCV-SOC curve which has a significant slope between ~20% and 40% SOC but a much flatter slope between ~40% and 80% SOC. Therefore, significant SOC imbalance can be maintained in a Li-ion cell with LFP electrode if the local SOC distribution is the range with flat OCV. Indeed, the flat OCV-SOC behavior was previously used to freeze non-uniform SOC distribution in Li-ion cells with LFP cathode for ex situ measurement [50]. Consisting of lithium anode and LFP cathode, the cell was charged to 50% SOC, and then the LFP electrode was removed from the cell to be scanned by synchrotron X-ray microdiffraction. Significant non-uniformity of charge distribution was preserved and observed in the LFP electrode. Both the relatively lower currents of sections 5–10 (local SOC > ~50%) in Fig. 2.13b, and the more significant recovery of SOC during rest for smaller (x/L) regions compared with the higher (x/L) sections, are also consistent with this argument.

These results imply that using nearly rested voltage as an indicator of cell internal balance may not be reliable. More specifically, the dV/dt tolerance to consider cell balancing may need to depend on the slope of the OCV-SOC curve in the region to which the voltage is resting. The recent battery excitation likely also plays a role in how appropriate a nearly rested voltage is in indicating a truly balanced battery. Further exploration of this phenomena is necessary.

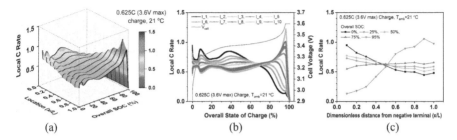

Fig. 2.15 Spatial-temporal current distribution during charging at 21 °C. (**a**) Variation of local currents with both location and overall SOC; (**b**) Variation of local currents and cell voltage with overall SOC; (**c**) Spatial current distribution at different overall SOC

2.2.2.7   Current Distribution During Charging

Figure 2.15 shows current distribution during charging at constant current of 1.5A, corresponding to 0.625C, followed by 3.6 V constant voltage charging until current drops to 0.02C. Due to change of charging current at constant voltage stage, local C rate is used to describe local currents. Similar to that during discharging, local current distribution is non-uniform during charging, especially at the beginning and end of constant current charging when cell voltage changes dramatically with SOC. Such non-uniform current distribution could dramatically exacerbate lithium plating during extreme fast charging that is actively pursued by EV industry. So further research on this would be particularly important. It is interesting to note that the trend of spatial current distribution is similar in both constant-current charging and discharging by comparing Figs. 2.5 and 2.15. At initial stages of charging or discharging, it is always segments closer to the negative terminal that experience higher local C rate. Since electrode materials degrade faster at higher C rate, it can be expected that such non-uniform current distribution could accelerate degradation of Li-ion cells.

2.2.2.8   Current Distribution During Partial Charging and Discharging

Li-ion cells in EV applications are not always fully charged or fully discharged. To investigate the effects of partial charging and discharging on current distribution, the experimental cell was cycled between 40% SOC and 60% SOC at 1C constant current three times. Before the cycling, the Li-ion cell was firstly fully charged, then discharged at C/5 to 60% SOC, followed by 1C discharge to 40% SOC, and then rest for 2 h. Figure 2.16 shows current distribution and cell voltage during such partial charging and discharging. Note that negative local C rate indicates the segments are charged. It can be seen that the current distribution behaviors are highly symmetrical during partial discharging and charging. Segments closer to the negative terminal always have higher local currents than those away from the negative terminal. For

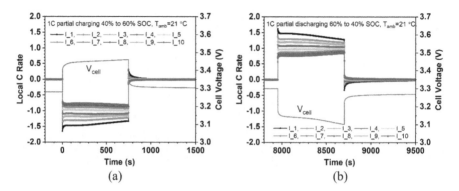

Fig. 2.16 Current distribution during partial charging and discharging cycling. (**a**) 1C charging from 40% SOC to 60% SOC; (**b**) 1C discharging from 60% SOC to 40% SOC

example, the local C rate of segment 1 ranges from 1.3 to 1.6, while that of segment 10 ranges only from 0.7 to 0.85. Even internal balancing currents are not enough to restore the non-uniformity. Such non-uniform current distribution and non-uniform material utilization would not only cause waste of material in locations far away from the tabs but also cause accelerated degradation of materials in locations closer to the tabs. Further investigation of how non-uniform current distribution influences degradation is needed to test this hypothesis.

2.2.2.9 Effects of Tab Configuration on Current Distribution and Usable Energy Density

Tab number and configurations have strong influence on performance of Li-ion batteries [14–15, 51]. But no quantitative relationship between cell energy density and current distribution was established experimentally in the literature. The EC Power-PSU team hypothesized that the effects of tab configurations on the energy density of Li-ion battery can be attributed to the effects on current distribution and experimentally validated the hypothesis.

Figure 2.17 shows schematically five tab configurations, all with the same negative tab but different positive tab configurations: ten positive tabs in parallel, five positive tabs in parallel, two positive tabs in parallel, one positive tab located counter to the negative tab, and one positive tab located at the same end of negative tab. When multiple positive electrode segments are connected in series, the additional resistance of positive tabs and shunt current sensors make the total resistance of positive current collector in this segmented cell much higher than that of a non-segmented cell. Therefore, the current distribution in this segment cell would be much more non-uniform than that in a non-segmented cell. It should be noted, however, that numerical models validated by the distribution data in this

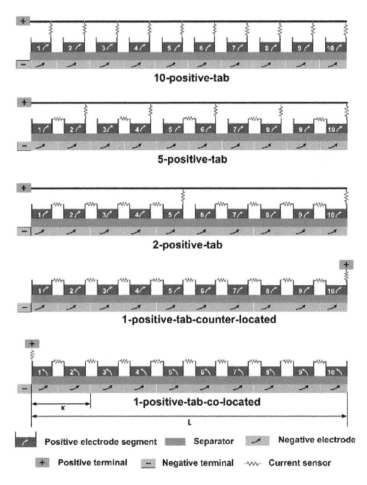

Fig. 2.17 Schematic of cell configurations with different positive tab numbers and locations. (Reused with permission from Ref. [28] Copyright 2013 The Electrochemical Society)

work can be used to explore various cases, including the case that both positive and negative electrodes are non-segmented.

Figure 2.18 shows overall performance of the experimental cell with different tab configurations. For convenience of comparison, the discharge capacity with 10-positive-tab configuration is used to calculate overall depth of discharge (DOD) for all the cases. Thus, the overall DOD in 10-tab case is set as 100%. As expected, tab number and location have significant effects on the overall performance of the experimental Li-ion cell. The cell voltage and discharge capacity are generally lower with fewer tabs, although the difference between 5-tab case and 10-tab case is very small. With same positive tab number, the 1-tab-co-located case has initially higher but then lower cell voltage than the 1-tab-counter-located case.

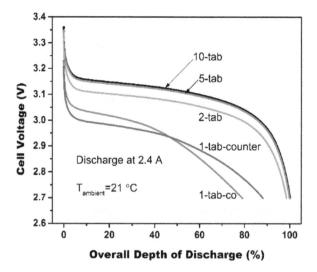

Fig. 2.18 Overall performance of the experimental Li-ion cell with different positive tab configurations. (Reused with permission from Ref. [28] Copyright 2013 The Electrochemical Society)

It is easy to understand that the generally lower cell performance with fewer tabs results directly from higher cell resistance. But the difference between 1-tab-coutner-located case and 1-tab-co-located is not straightforward from overall performance comparison. Instead, it can be clearly explained by current distribution results in the two cases.

Figure 2.19 shows current distribution in the 1-tab-countered-located case. It can be seen that local current in segment 10 has the highest initial value and becomes

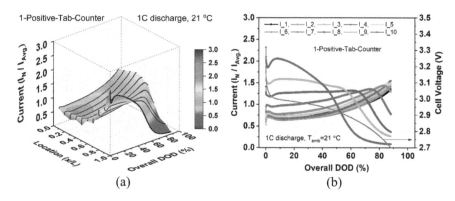

(a)　　　　　　　　　　　　　　　　　　　　　　(b)

Fig. 2.19 Current distribution during discharge in 1-tab-counter-located configuration. (**a**) Variation of local currents with both location and overall SOC; (**b**) Variation of local currents and cell voltage with overall SOC. (Adapted with permission from Ref. [28] Copyright 2013 The Electrochemical Society)

lowest near the end of discharge. The non-uniform current distribution at beginning of discharge can be attributed to the effects of current collector resistance, especially the positive side. Local currents in segments 1–9 must flow through downstream segments, and the positive side resistance of downstream segments tends to make upstream local currents lower and downstream local currents higher. In this experimental cell, the resistance of current collector on positive side consists of segment aluminum foil resistance, segment tab resistance, and current sensor resistance, making positive side resistance higher than the negative side. As a result, the effects of positive side resistance on current distribution are dominant, and segment 10 generates the highest initial local current. Nevertheless, the counteracting effects of negative side current collector resistance can be still observed from the initial current distribution, in which segments 1 and 2 generate higher local currents than segment 3. Higher initial local current leads to faster depletion of active material. So the local current in segment 10, which has highest initial value, quickly depletes and becomes lowest. The depletion then propagates upstream as discharge proceeds.

Figure 2.20 shows current distribution in the 1-tab-co-located case. From comparison with Fig. 2.19, it can be seen that the current distribution in this case is almost opposite to that in the 1-tab-counter-located case, and the current distribution is even less uniform. Difference of current distribution in the two cases can be attributed to how the effects of positive current collector resistance and negative current collector resistance interact. As discussed above, in the 1-tab-counter-located case, the effects of positive side resistance and negative side resistance are counteracting. In the 1-tab-co-located case, however, the effects are synergizing, both favoring higher local currents in upstream segments. Therefore, in the 1-tab-co-locate case, the local current in segment 1 is exceptionally high and that in segment 10 exceptionally low during initial period of discharge. As discharge

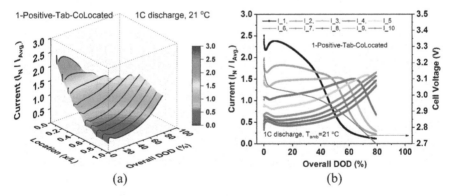

(a)   (b)

Fig. 2.20 Current distribution during discharge in 1-tab-co-located case. (**a**) Variation of local currents with both location and overall SOC; (**b**) Variation of local currents and cell voltage with overall SOC. (Adapted with permission from Ref. [28] Copyright 2013 The Electrochemical Society)

proceeds, the stored energy in segment 1 is more quickly depleted, and the local current quickly becomes the lowest.

With detailed results of current distribution in the 1-tab-co-located case and the 1-tab-counter-located case, the difference of overall cell performance shown in Fig. 2.17 can be clearly explained. Initially, a great portion of total current is generated in upstream segments in the 1-tab-co-located case, while the situation is opposite in the 1-tab-counter-located case. So the average travel distance of local currents through the positive and negative and current collectors is much shorter in the 1-tab-co-located case. Consequently, the average resistance is lower. With the battery cell discharged at the same total current, the voltage drop is lower in 1-tab-co-located case, resulting in increased utilization. As discharge proceeds, however, stored energy in upstream segments is depleted quickly due to very high initial local currents. As the depletion moves downstream, reactions in downstream segments are accelerated to keep the total current constant. As a result, the average travel distance of the total current becomes longer, with the average resistance and voltage drop becoming higher. The variation of current distribution, the average travel distance of total current, and the average resistance is essentially opposite in the 1-tab-counter-located case. Therefore, the difference of cell voltage between the two cases becomes smaller as the discharge proceeds. Eventually the cell voltage in the 1-tab-co-located case becomes equal to, and then lower than, that in the 1-tab-counter-located case.

The very non-uniform current distribution in 1-tab-co-located case suggests two potential problems. First, energy storage materials in downstream segments are significantly underutilized, leading to reduced energy density, an undesired waste. Second, electrodes closer to tabs produce higher local currents (higher C rate) than average and experience faster depletion during discharge, even overdischarge of active materials. Previous studies showed that a Li-ion cell degrades faster for higher C rate and higher DOD [52–53], so electrode materials near tabs are likely to degrade faster, eventually causing the whole cell to degrade faster than a cell with uniform current distribution. Therefore, improving current distribution in Li-ion cells through design optimization is beneficial for energy density and durability, both of which are major challenges for electric vehicle applications. Correlating current distribution non-uniformity with degradation through in situ measurement of current distribution in aging tests can provide more insights about large-format Li-ion cell degradation. Further research efforts along these directions are warranted.

The non-uniform utilization of active materials in 1-tab-co-located case is further demonstrated in Fig. 2.21, in which DOD distributions at cutoff in different cases are compared. It can be seen that the DOD distribution in the 10-tab and 5-tab cases are very similar and rather uniform, with only downstream segments slightly underutilized. In comparison, some segments are much less utilized in other cases depending on the tab configuration, particularly in the 1-tab-co-located and 1-tab-counter-located cases. It is interesting to note that local DOD of upstream segments in the 1-tab-co-located case is very similar to the 10-tab case and 5-tab case, while that of downstream segments is much lower. This clearly shows the effects of current distribution on energy utilization of Li-ion cell.

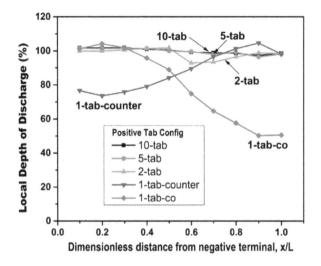

Fig. 2.21 Comparison of local DOD distribution at cutoff with different tab configurations. (Reused with permission from Ref. [28] Copyright 2013 The Electrochemical Society)

The very similar overall performance, current distribution, and DOD distribution in 5-tab case and 10-tab case suggests that excessive tab number helps little. Actually, excessive tabs increase the complexity and cost of battery cell manufacturing, which is not desired.

2.2.2.10 Correlation Between Energy Density and Current Distribution Non-uniformity

The current distribution and DOD distribution results in Fig. 2.21 indicate that non-uniform current distribution lead to underutilization of active materials and lower energy density. To establish the correlation between non-uniform current distribution and energy density more clearly, discharge energy is plotted as a function of current distribution non-uniformity factor [15] in Fig. 2.22. Discharge energy is normalized by that during C/5 discharge with 10-tab configuration, which is assumed to be the maximum energy available from the experimental cell. It can be seen that the normalized energy decreases almost linearly with the increase of current distribution non-uniformity factor. It demonstrates a significant effect of current distribution non-uniformity on energy density and the importance of improving current distribution uniformity in energy-dense Li-ion batteries for vehicle energy storage.

The experimental correlation between usable energy density and current distribution non-uniformity results was later used by Zhao et al. [15] to validate numerical models. As shown in Fig. 2.23, the model results of 40 Ah Li-ion cells were superimposed with the experimental data of the 2.4 Ah LFP/graphite cell. The

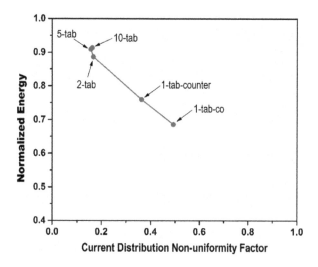

Fig. 2.22 Normalized discharge energy vs. current distribution non-uniformity factor. (Reused with permission from Ref. [28] Copyright 2013 The Electrochemical Society)

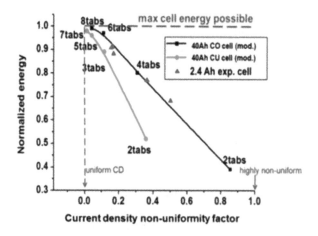

Fig. 2.23 Comparison of modeling results for 40 Ah cells by Zhao et al. [15] with experimental data for 2.4 Ah segmented Li-ion cell. (Reused with permission from Ref. [15] Copyright 2014 Elsevier B.V)

energy density was for 1 C discharge process and was normalized by the coin cell energy density, i.e., the maximum achievable energy density with the battery materials used. Despite differences in size and cathode material, the trend and magnitude of cell energy density vs. current density non-uniformity are clearly consistent between the model predictions and experiments. More quantitative comparisons using the NMC chemistry and the exact cell size need to be carried out in the future.

2.3   Indirect Diagnosis of Current Distribution Through Local Potential Measurement

As schematically shown in Fig. 2.1, when current flows in copper foil and aluminum foil current collectors, there will be electric potential change along the length of current collectors due to ohmic resistance. By measuring distribution of local potential along the foils, it is possible to estimate local currents through numerical models. A team at Technical University of Munich (TUM) and their collaborators [31–33] took this approach, firstly using a modified commercial cylindrical cell [31–32] and then using a specially developed single-layered pouch cell [33]. The team's work is reviewed in this section.

2.3.1   Experimental Method Using Modified Commercial Cylindrical Cells

Figures 2.24 and 2.25 show schematic and pictures of cylindrical cell before and after modification by Osswald et al. [31]. Before modification, there are four parallel connected tabs on negative electrode (A1–A4) and four parallel connected tabs on positive electrode (C1–C4). External current is applied to the cell through all the tabs, and only one voltage can be measured. After modification, the four tabs on each electrode are separated. External current is applied only through one pair of tables (A1 and C1) in the study, creating a co-located tab configuration which would create more non-uniform current distribution than other tab configurations [15, 28]. The modification allows for voltage measurements at four different positions of the electrode, which are used to validate models and simulation of local current densities [32].

The cylindrical cells used for modification purposes are commercially available 26650 format cells with a nominal capacity of 2.5 Ah and an average discharge voltage of 3.3 V. The cathode has an overall length of 1.69 m, the anode 1.75 m, and each electrode has four almost equidistant current collecting tabs. The aluminum current collector is 20 μm thick, and the copper current collector is 13 μm thick. Lithium iron phosphate is used for the cathode's active material, and graphite is used for the anode. More experimental details are also reported [31].

2.3.2   Results from Modified Commercial Cylindrical Cell

Figure 2.26 shows measured and simulated voltages at each of the four tab pairs of the modified cell during 0.5C, 1C, and 2C discharge at room temperature. It can be seen that local voltage measured at tab 1 is significantly lower than those at other locations and the difference increased at higher C rate. The results clearly

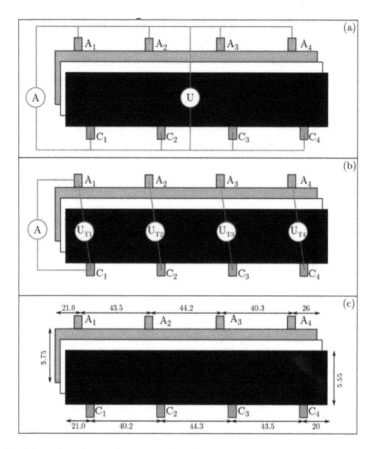

Fig. 2.24 Schematic and operation mode of experimental cell by Osswald et al. [31]. (**a**) Before modification all four tabs on each electrode (A1–A4 on negative electrode and C1–C4 on positive electrode) are connected in parallel and applied external current; (**b**) After modification tabs are separated, allowing voltage measurements at four different positions, and only outermost pair of tables (A1 and C1) is applied external current; (**c**) Locations of tabs. (Reused with permission from Ref. [31] Copyright 2015 The Electrochemical Society)

demonstrated the effects of current collector resistance on local voltage distribution, which would influence current distribution. Note that when tab 1 voltage reached cutoff value of 2.4 V at the end of 2C discharge, other tab voltages are still 2.8 V and above, suggesting underutilization of capacity in regions farther away from terminals [32], consistent with the EC Power-PSU team's findings on co-located tab configuration [15, 28].

Figure 2.27 shows simulated current density in electrodes at each of the four tab pairs corresponding to the voltage results in Fig. 2.25. It can be seen that the trend of current distribution is consistent and similar to that measured by the EC Power-PSU team with co-located tab configuration. Local current densities in region closest to terminals are highest at beginning of discharge but quickly decrease due to depletion

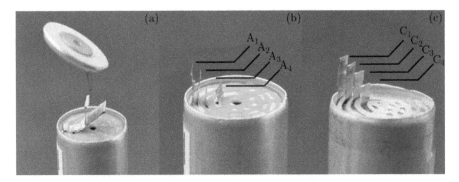

Fig. 2.25 Pictures of experimental cell by Osswald et al. [31] (**a**) Opened on the negative electrode with four tabs connected; (**b**) Opened on the negative end with current tabs separated; (**c**) Opened on the positive electrode with current tabs separated. (Reused with permission from Ref. [31] Copyright 2015 The Electrochemical Society)

of local SOC. Then local current densities in regions further away from the terminals increase correspondingly and then also decrease due to depletion of SOC. Such a distinct trend is governed by the same mechanism, i.e., nonlinear decrease of local open circuit voltage with SOC, primarily from the graphite anode.

2.3.3   Experimental Method Using Single-Layered Pouch Li-Ion Cell

As shown in Fig. 2.28, Erhard et al. [33] developed a single-layered pouch cell with ten tab pairs at each edge for potential distribution measurement. There is also a reference tab pair at the left and right side of the cell to be connected to battery tester. The current collectors are 500 mm long and 100 mm wide. The aluminum current collector is 20 μm thick, and the copper current collector is 18 μm thick. Compared with the segmented pouch cell by the EC Power-PSU team and the cylindrical cell used by the same TUM team, the length of current collector in this work is much shorter but still comparable to large-format pouch cells in which electrodes are stacked instead of being wounded. The copper foil in this work is thicker than normally used which could influence current distribution behaviors. A unique feature of this experimental cell is single-layered structure, which can dramatically reduce effects of non-uniform temperature distribution and heat generation on current distribution as experienced in the large cylindrical cell [32].

The cell comprises a $LiNi_{0.33}Mn_{0.33}Co_{0.33}O_2$ (NMC) cathode and a graphite anode with nominal capacity of 0.8 Ah. NMC is increasingly used in EV applications, and its open circuit potential behavior is different from LFP cathode, which would influence current distribution behaviors.

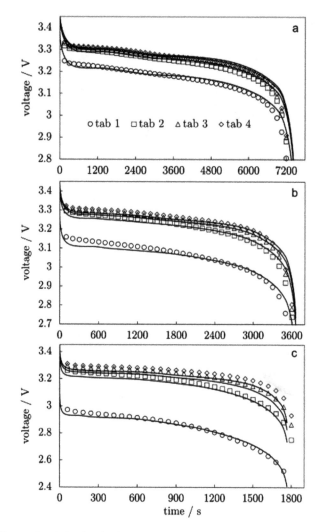

Fig. 2.26 Measured and simulated voltages across each of the four tab pairs during (**a**) 0.5C, (**b**) 1C, and (**c**) 2C discharge at room temperature. (Reused with permission from Ref. [32] Copyright 2015 The Electrochemical Society)

Figure 2.29 shows schematically the team's experimental setup [33]. In their study, the current from the battery tester was only applied at one of the two reference tab pairs, creating a co-located tab configuration similar to that in their work on cylindrical cell. This tab configuration would create more non-uniform current distribution than other configurations, and it is not advised for actual application [15, 28]. A climate chamber was used to control ambient temperature.

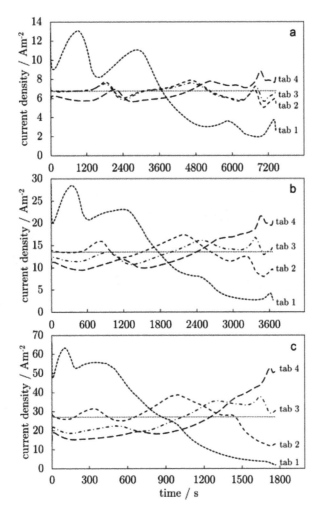

Fig. 2.27 Simulated current density in electrodes at each of the four tab pairs during (**a**) 0.5C, (**b**) 1C, and (**c**) 2C discharge at room temperature. (Reused with permission from Ref. [32] Copyright 2015 The Electrochemical Society)

2.3.4   Results from Single-Layered Pouch Li-Ion Cell

Figure 2.30 shows experimental and simulation results during 0.5C, 1C, and 2C discharge at a temperature of 25 °C by Erhard et al. [33]. It can be seen that the wavy current distribution behaviors are similar to the results reported by the EC Power-PSU team [27], suggesting similar controlling mechanism of current distributions in Li-ion cells despite different electrode materials. Local measurement point #01 is much closer to the terminals (reference tabs) than measurement point #10, so the resistance is lower, causing higher local current at point #01 than that at #10 at initial

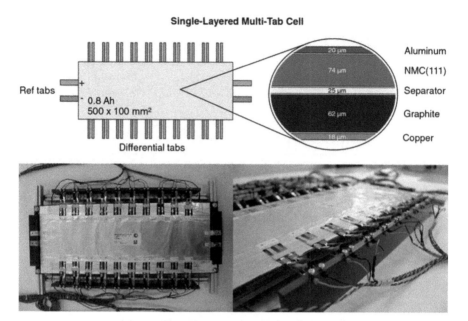

Fig. 2.28 Schematic representation and pictures of a single-layered pouch Li-ion cell for potential distribution measurement by Erhard et al. [33]. (Reused under the terms of the Creative Commons Attribution 4.0 License (CC BY) from Ref. [33] Copyright 2017 The Authors. Published by ECS)

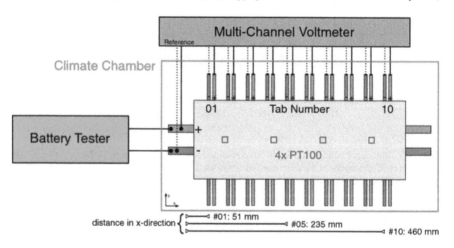

Fig. 2.29 Schematic representation of experimental setup for potential distribution measurement by Erhard et al. [33]. (Reused under the terms of the Creative Commons Attribution 4.0 License (CC BY) from Ref. [33] Copyright 2017 The Authors. Published by ECS)

stages of discharge. As discharging proceeds, local SOC and therefore local OCV would cause the current distribution to reverse near the end of discharge. The local current curves are less wavy during 2C discharge than during 0.5C discharge, also

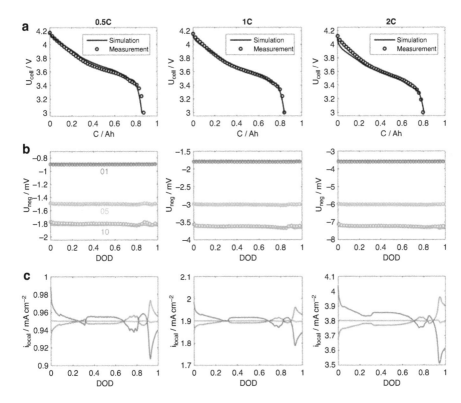

Fig. 2.30 Experimental and simulation results during 0.5C, 1C, and 2C discharge at a temperature of 25 °C by Erhard et al. [33]. (**a**) Measured (symbols) and simulated (solid lines) cell voltage; (**b**) Measured and simulated potentials at local measurement points #01, #05, and #10; (**c**) Simulated current densities at the corresponding points. (Reused under the terms of the Creative Commons Attribution 4.0 License (CC BY) from Ref. [33] Copyright 2017 The Authors. Published by ECS)

consistent with the results reported by the EC Power-PSU team [27] on the effects of discharging C rate.

Figure 2.31 shows experimental and simulation results during 1C discharge at temperatures of 5 °C, 25 °C and 40 °C by Erhard et al. [33]. Current distributions are more non-uniform and wavy at higher temperatures which could be attributed to the change of reaction kinetics and cell resistance at different temperatures.

Erhard et al. [33] also simulated local SOC distribution during discharge. Figure 2.32 shows simulated difference in SOC between tab #01 and tab #10 during 0.5C, 1C, and 2C discharge at 25 °C. As expected, the difference initially increased due to difference in local currents but eventually reduced near the end of discharge due to reverse of current distribution trend. The difference is more significant at higher C rate and higher temperature, also corresponding to the behaviors of current distribution. It is interesting to note that the SOC difference at end of 2C discharge is much smaller than that reported by the EC Power-PSU team [27], which can be

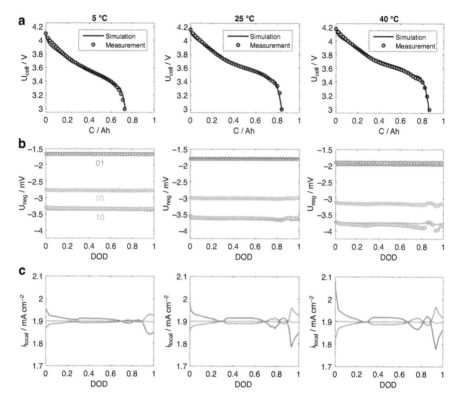

Fig. 2.31 Experimental and simulation results during 1C discharge at temperatures of 5 °C, 25 °C and 40 °C by Erhard et al. [33]. (**a**) Measured (symbols) and simulated (solid lines) cell voltage; (**b**) Measured and simulated potentials at local measurement points #01, #05, and #10; (**c**) Simulated current densities at the corresponding points. (Reused under the terms of the Creative Commons Attribution 4.0 License (CC BY) from Ref. [33] Copyright 2017 The Authors. Published by ECS)

attributed to difference between NMC cathode and LFP cathode in dependence of open circuit potential on SOC. The difference would be more dramatic at higher C rate and is worth further investigation.

2.4   Noninvasive Diagnosis of Current Distribution Using Magnetic Resonance Imaging

A magnetic field is generated as electric current flows in a Li-ion cell. By mapping the generated magnetic field using magnetic resonance imaging techniques, it is possible to noninvasively detect current flow inside Li-ion cells. A team at New York University recently demonstrated this concept [34]. In an earlier work, the

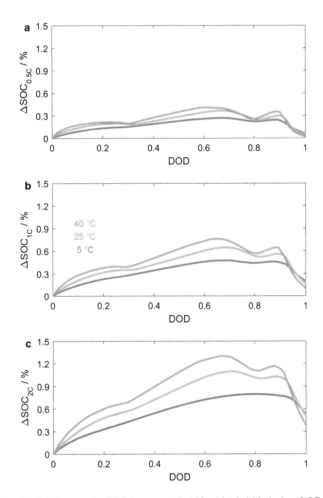

Fig. 2.32 Simulated difference in SOC between tab #01 and tab #10 during 0.5C, 1C, and 2C discharge at 25 °C [33]. (Reused under the terms of the Creative Commons Attribution 4.0 License (CC BY) from Ref. [33] Copyright 2017 The Authors. Published by ECS)

team and their collaborators demonstrated the use of magnetic resonance imaging to noninvasively determine Li-ion cell state of charge [54].

2.4.1 Measurement Method Using Magnetic Resonance Imaging

Figure 2.33a shows an illustration of the experimental arrangement by Mohammadi et al. [34]. A pouch-type Li-ion cell is placed in a customer-designed holder

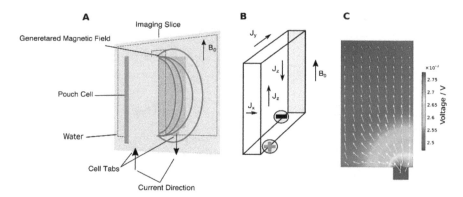

Fig. 2.33 Illustration of the experimental arrangement by Mohammadi et al. [34]. (**a**) Cell position and orientation of imaging slice, with dashed rectangle indicating detected volume; (**b**) Directions of current flow and cell orientation; (**c**) Calculated current and voltage distribution on the positive current collector during charging. (Reused with permission from Ref. [34] Copyright 2019 Elsevier Inc.)

containing water compartments that sandwiching the Li-ion cell. 15 mM $CuSO_4$ water solution is filled in the compartments as detection medium. A strong external static magnetic field B_0 is applied along z direction (from cell end with tabs to the opposite end). Then the magnetic field generated by current flowing along y direction (J_y in Fig. 2.33b) is detected by comparing magnetic field maps with current flow and at rest. As shown in Fig. 2.33c, the authors also calculated current distribution in positive current collector during charging, but the calculation seemed to be based on a simplified electric model and not based on measured magnetic field data.

2.4.2 Results from Magnetic Resonance Imaging

Mohammadi et al. [34] demonstrated the technique using Li-ion pouch cells with capacity of 250 mAh. Each cell was composed of five double-sided cathodes and six double-sided anodes. The active materials in cathode and anode were $Li_{1.02}Ni_{0.50}Mn_{0.29}Co_{0.19}O_2$ and graphite, respectively. The electrolyte was 1.2 M $LiPF_6$ in EC:DMC 3:7 (%wt). Figure 2.34 shows magnetic field maps and histograms during discharge and charge at the imaging slice indicated by the dashed rectangle in Fig. 2.33a. Negative current indicates discharging and positive current indicates charging. The magnetic field maps were obtained after subtracting a reference image from resting period and a constant background field. Red color indicates positive change of magnetic field, and blue color indicates negative change of magnetic field. The discharge or charge current was 125 mA, corresponding to 0.5C. The results show the influence of internal current flow on the magnetic field.

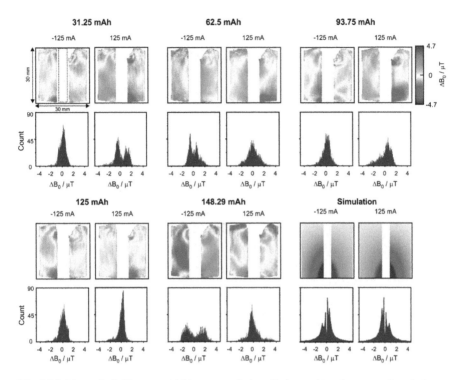

Fig. 2.34 Magnetic field maps and histograms during discharge (negative current) and charge (positive current) at different depth of discharge as indicated. The battery tabs in these measurements are located at the bottom of each map. The imaging dimensions and the location of the cell (dotted rectangle) are indicated in the top left image. (Reused with permission from Ref. [34] Copyright 2019 Elsevier Inc.)

It can be seen that the changes of magnetic field depend on direction of current and depth of discharge, which could be used to detect existence of current flow or even local current distribution. However, quantified current distribution from measured magnetic field was not reported. The study also examined correlation of magnetic field with applied current and found significant deviation from linearity. Further work would be needed to quantify current distribution from magnetic resonance imaging.

2.5 Summary and Future Work

As reviewed above, non-uniform distribution in large-format Li-ion cells has significant negative influences in power performance, usable energy density, durability, and safety, which are critical to EV applications. Novel techniques, including segmented cells, cells with embedded potential sensors, and noninvasive imaging, have been demonstrated as effective tools for in situ diagnosis of current distribution. Moreover, insights on Li-ion cells were obtained through spatial-temporal distribution of current densities that could not be obtained from overall cell performance characterization. In particular, the non-uniform and wavy distribution of local currents are experimentally found. The effects of non-uniform current distribution on underutilization of active materials, internal balancing current, and its influence on SOC redistribution and cell voltage recovery were observed. The effects of tab configuration on current distribution and thus usable energy density were revealed. Experimental spatial-temporal data are also important in validating multidimensional electrochemical-thermal coupled models that are increasingly used for Li-ion battery development.

With the widespread adoption of electric vehicles around the globe, which increasingly demands for Li-ion batteries with higher material utilization, further efforts are needed in diagnosing current distribution in large-format Li-ion cells for insights to develop faster rechargeable, more energy dense, safer, and more durable Li-ion batteries. Here we propose a few topics that are particularly worth investigating from EV battery perspective.

The first suggested topic is measurement of current distribution under extreme conditions such as extreme fast charging, extreme thermal conditions (as low as $-40\ °C$ and as high as $80\ °C$), and safety abuse conditions (mechanical puncture, crash, or internal short circuit). Lithium plating is a critical challenge for fast charging [55], and it can be exacerbated by non-uniform current distribution. Under extreme low temperatures and rapid heating process, the internal resistance of Li-ion cells changes dramatically, which would cause dramatic change of current distribution. Under safety abuse conditions, the current distribution would be extremely non-uniform and dynamically changing [25, 56]. In situ measurement of current distribution under these extreme yet important conditions would help achieve better understanding and development of Li-ion cells that perform better and more safely.

The second suggested topic is measurement of current distribution in Li-ion cells during long-term or accelerated aging. A recent study by Cavalheiro et al. [37] showed that current distribution in a small module consisting of parallel-connected Li-ion cells greatly varied during aging and contributed to accelerated aging. A similar mechanism is expected to exist at the cell level and worth further investigation for large-size cells. The effects of cooling conditions, cell constraint, and compression on current density distribution and failure are also necessary.

The third suggested topic is measurement of current distribution in Li-ion cells with emerging materials, such as low cobalt cathode, silicon-based anode, or solid-state electrolytes. Current distribution behaviors of cells with these materials could be very different as implied by the comparison of studies with LFP cathode [27] and NMC cathode [33]. In particular, silicon anode [57] and low cobalt NMC cathode [58] degrade faster than conventional electrode materials. The interaction between degradation and current distribution is worth investigation.

The last but not the least suggested topic is development of new methods for current distribution diagnosis in large-format EV Li-ion cells. While the reported methods revealed many interesting phenomena, each has its own disadvantages. The segmented cell method enables direct measurement of current distribution, but it requires significant modification of Li-ion cells. Indirect measurement with embedded potential sensors requires less modification, but calculating current distribution from potential distribution requires thermal-electrochemical modeling. Noninvasive diagnosis through magnetic resonance does not require any modification of experimental cells, but quantification of current distribution from magnetic field map and its application to extreme conditions could be challenging. Moreover, the reported current distributions are primarily one dimensional. It would be necessary to extend the measurement to two dimensional or three dimensional for comprehensive understanding of the behaviors in a large-format EV Li-ion cells with capacity of more than 100 Ah. It would be also valuable to integrate the current distribution diagnosis with other in situ diagnostic tools, such as in situ temperature distribution measurement [29, 56, 59–61], in situ neutron imaging [62], and X-ray methods [63–64].

References

1. Whittingham MS (2012) History, evolution, and future status of energy storage. Proc IEEE 100(Special Centennial Issue):1518–1534
2. Winter M, Barnett B, Xu K (2018) Before Li ion batteries. Chem Rev 118(23):11433–11456
3. Bloomberg New Energy Finance (2020) Electric vehicle outlook 2020. https://about.bnef.com/electric-vehicle-outlook/
4. DOE Office of Science (2017) Report of basic research needs workshop on next generation electric energy storage. https://science.energy.gov/~/media/bes/pdf/reports/2017/BRN_NGEES_rpt.pdf
5. DOE (2020) 2020 Annual merit review report. https://www.energy.gov/eere/vehicles/downloads/2020-annual-merit-review-report
6. Ahmed S, Bloom I, Jansen AN, Tanim T, Dufek EJ, Pesaran A, Burnham A, Carlson RB, Dias F, Hardy K, Keyser M, Kreuzer C, Markel A, Meintz A, Michelbacher C, Mohanpurkar M, Nelson PA, Robertson DC, Scoffield D, Shirk M, Stephens T, Vijayagopal R, Zhang J (2017) Enabling fast charging – a battery technology gap assessment. J Power Sources 367(Supplement C):250–262
7. Wang C-Y, Xu T, Ge S, Zhang G, Yang X-G, Ji Y (2016) A fast rechargeable lithium-ion battery at subfreezing temperatures. J Electrochem Soc 163(9):A1944–A1950
8. Yang X-G, Zhang G, Ge S, Wang C-Y (2018) Fast charging of lithium-ion batteries at all temperatures. Proc Natl Acad Sci 115(28):7266–7271
9. Yang X-G, Liu T, Gao Y, Ge S, Leng Y, Wang D, Wang C-Y (2019) Asymmetric temperature modulation for extreme fast charging of lithium-ion batteries. Joule 3(12):3002–3019
10. Wang C-Y, Zhang G, Ge S, Xu T, Ji Y, Yang X-G, Leng Y (2016) Lithium-ion battery structure that self-heats at low temperatures. Nature 529(7587):515–518
11. Feng X, Ouyang M, Liu X, Lu L, Xia Y, He X Thermal runaway mechanism of lithium ion battery for electric vehicles: a review. Energy Storage Mater 2018, 10:246–267
12. Ruiz V, Pfrang A, Kriston A, Omar N, Van den Bossche P, Boon-Brett L (2018) A review of international abuse testing standards and regulations for lithium ion batteries in electric and hybrid electric vehicles. Renew Sust Energ Rev 81:1427–1452
13. Sun P, Bisschop R, Niu H, Huang X (2020) A review of battery fires in electric vehicles. Fire Technol 56(4):1361–1410

14. Lee K-J, Smith K, Pesaran A, Kim G-H (2013) Three dimensional thermal-, electrical-, and electrochemical-coupled model for cylindrical wound large format lithium-ion batteries. J Power Sources 241:20–32
15. Zhao W, Luo G, Wang C-Y (2014) Effect of tab design on large-format Li-ion cell performance. J Power Sources 257:70–79
16. Doyle M, Fuller TF, Newman J (1993) Modeling of galvanostatic charge and discharge of the lithium/polymer/insertion cell. J Electrochem Soc 140(6):1526–1533
17. Smith K, Wang C-Y (2006) Power and thermal characterization of a lithium-ion battery pack for hybrid-electric vehicles. J Power Sources 160:662–673
18. Smith K, Wang C-Y (2006) Solid-state diffusion limitations on pulse operation of a lithium ion cell for hybrid electric vehicles. J Power Sources 161:628–639
19. Ramadesigan V, Methekar RN, Latinwo F, Braatz RD, Subramanian VR (2010) Optimal porosity distribution for minimized ohmic drop across a porous electrode. J Electrochem Soc 157(12):A1328–A1334
20. Santhanagopalan S, Guo Q, Ramadass P, White RE (2006) Review of models for predicting the cycling performance of lithium ion batteries. J Power Sources 156(2):620–628
21. Ramadesigan V, Northrop PWC, De S, Santhanagopalan S, Braatz RD, Subramanian VR (2012) Modeling and simulation of lithium-ion batteries from a systems engineering perspective. J Electrochem Soc 159(3):R31–R45
22. Fang W, Kwon OJ, Wang C-Y (2010) Electrochemical-thermal modeling of automotive Li-ion batteries and experimental validation using a three-electrode cell. Int J Energy Res 34:107–115
23. Luo G, Wang CY (2012) A multidimensional, electrochemical-thermal coupled lithium-ion battery model, Chapter 7. In: Yuan X, Liu H, Zhang J (eds) Lithium-ion batteries. CRC Press, Boca Raton
24. Zhao W, Luo G, Wang C-Y (2015) Modeling internal shorting process in large-format Li-ion cells. J Electrochem Soc 162(7):A1352–A1364
25. Zhao W, Luo G, Wang C-Y (2015) Modeling nail penetration process in large-format Li-ion cells. J Electrochem Soc 162(1):A207–A217
26. Yang XG, Zhang G, Wang CY (2016) Computational design and refinement of self-heating lithium ion batteries. J Power Sources 328:203–211
27. Zhang G, Shaffer CE, Wang C-Y, Rahn CD (2013) In-situ measurement of current distribution in a Li-ion cell. J Electrochem Soc 160(4):A610–A615
28. Zhang G, Shaffer CE, Wang CY, Rahn CD (2013) Effects of non-uniform current distribution on energy density of Li-ion cells. J Electrochem Soc 160(11):A2299–A2305
29. Zhang G, Cao L, Ge S, Wang C-Y, Shaffer CE, Rahn CD (2014) In situ measurement of radial temperature distributions in cylindrical Li-ion cells. J Electrochem Soc 161(10):A1499–A1507
30. Zhang G, Cao L, Ge S, Wang C-Y, Shaffer CE, Rahn CD (2015) Reaction temperature sensing (RTS)-based control for Li-ion battery safety. Sci Rep 5:18237
31. Osswald PJ, Erhard SV, Wilhelm J, Hoster HE, Jossen A (2015) Simulation and measurement of local potentials of modified commercial cylindrical cells: I: cell preparation and measurements. J Electrochem Soc 162(10):A2099–A2105
32. Erhard SV, Osswald PJ, Wilhelm J, Rheinfeld A, Kosch S, Jossen A (2015) Simulation and measurement of local potentials of modified commercial cylindrical cells: II: multi-dimensional modeling and validation. J Electrochem Soc 162(14):A2707–A2719
33. Erhard SV, Osswald PJ, Keil P, Höffer E, Haug M, Noel A, Wilhelm J, Rieger B, Schmidt K, Kosch S, Kindermann FM, Spingler F, Kloust H, Thoennessen T, Rheinfeld A, Jossen A (2017) Simulation and measurement of the current density distribution in lithium-ion batteries by a multi-tab cell approach. J Electrochem Soc 164(1):A6324–A6333
34. Mohammadi M, Silletta EV, Ilott AJ, Jerschow A (2019) Diagnosing current distributions in batteries with magnetic resonance imaging. J Magn Reson 309:106601
35. Lai X, Jin C, Yi W, Han X, Feng X, Zheng Y, Ouyang M (2021) Mechanism, modeling, detection, and prevention of the internal short circuit in lithium-ion batteries: Recent advances and perspectives. Energy Storage Mater 35:470–499

36. Schindler M, Durdel A, Sturm J, Jocher P, Jossen A (2020) On the impact of internal cross-linking and connection properties on the current distribution in lithium-ion battery modules. J Electrochem Soc 167(12):120542
37. Cavalheiro, G. M.; Iriyama, T.; Nelson, G. J.; Huang, S.; Zhang, G. (2020) Effects of nonuniform temperature distribution on degradation of lithium-ion batteries. J Electrochem Energy Convers Storage 17(2):021101
38. Hu Y, Iwata GZ, Mohammadi M, Silletta EV, Wickenbrock A, Blanchard JW, Budker D, Jerschow A (2020) Sensitive magnetometry reveals inhomogeneities in charge storage and weak transient internal currents in Li-ion cells. Proc Natl Acad Sci 117(20):10667–10672
39. Ng S-H, La Mantia F, Novak P (2009) A multiple working electrode for electrochemical cells: a tool for current density distribution studies. Angew Chem Int Ed 48:528–532
40. Kindermann FM, Osswald PJ, Klink S, Ehlert G, Schuster J, Noel A, Erhard SV, Schuhmann W, Jossen A (2017) Measurements of lithium-ion concentration equilibration processes inside graphite electrodes. J Power Sources 342:638–643
41. Finegan DP, Quinn A, Wragg DS, Colclasure AM, Lu X, Tan C, Heenan TMM, Jervis R, Brett DJL, Das S, Gao T, Cogswell DA, Bazant MZ, Di Michiel M, Checchia S, Shearing PR, Smith K (2020) Spatial dynamics of lithiation and lithium plating during high-rate operation of graphite electrodes. Energy Environ Sci 13(8):2570–2584
42. Teufl T, Pritzl D, Solchenbach S, Gasteiger HA, Mendez MA (2019) Editors' choice—state of charge dependent resistance build-up in Li- and Mn-rich layered oxides during lithium extraction and insertion. J Electrochem Soc 166(6):A1275–A1284
43. Ohzuku T, Iwakoshi Y, Sawai K (1993) Formation of lithium-graphite intercalation compounds in nonaqueous electrolytes and their application as a negative electrode for a lithium ion (shuttlecock) cell. J Electrochem Soc 140(9):2490–2498
44. Yamada A, Koizumi H, Nishimura SI, Sonoyama N, Kanno R, Yonemura M, Nakamura T, Kobayashi Y (2006) Room-temperature miscibility gap in LixFePO4. Nat Mater 5(5):357–360
45. Bard AJ, Faulkner LR (2001) Electrochemical methods: fundamentals and applications, 2nd edn. Wiley, New York
46. Zhang Y, Wang C-Y (2009) Cycle-life characterization of automotive lithium-ion batteries with LiNiO2 cathode. J Electrochem Soc 156:A527–A535
47. Liao, L.; Zuo, P.; Ma, Y.; Chen, X.; An, Y.; Gao, Y.; Yin, G., Effects of temperature on charge/discharge behaviors of LiFePO4 cathode for Li-ion batteries. Electrochim Acta 2012, 60:269–273
48. Valoen LO, Reimers JN (2005) Transport properties of $LiPF_6$-based Li-ion battery electrolytes. J Electrochem Soc 152(5):A882–A891
49. Laughton MA, Warne DF (2003) Electrical engineer's reference book, 16th edn. Elsevier, Oxford
50. Liu J, Kunz M, Chen K, Tamura N, Richardson TJ (2010) Visualization of charge distribution in a lithium battery electrode. J Phys Chem Lett 1:2120–2123
51. Chen Y-S, Chang K-H, Hu C-C, Cheng T-T (2010) Performance comparisons and resistance modeling for multi-segment electrode designs of power-oriented lithium-ion batteries. Electrochim Acta 55:6433–6439
52. Vetter J, Novák P, Wagner MR, Veit C, Möller KC, Besenhard JO, Winter M, Wohlfahrt-Mehrens M, Vogler C, Hammouche A (2005) Ageing mechanisms in lithium-ion batteries. J Power Sources 147(1–2):269–281
53. Omar N, Monem MA, Firouz Y, Salminen J, Smekens J, Hegazy O, Gaulous H, Mulder G, Van den Bossche P, Coosemans T, Van Mierlo J (2014) Lithium iron phosphate based battery – assessment of the aging parameters and development of cycle life model. Appl Energy 113:1575–1585
54. Ilott AJ, Mohammadi M, Schauerman CM, Ganter MJ, Jerschow A (2018) Rechargeable lithium-ion cell state of charge and defect detection by in-situ inside-out magnetic resonance imaging. Nat Commun 9(1):1776

55. Yang X-G, Ge S, Liu T, Leng Y, Wang C-Y (2018) A look into the voltage plateau signal for detection and quantification of lithium plating in lithium-ion cells. J Power Sources 395:251–261

56. Huang S, Du X, Richter M, Ford J, Cavalheiro GM, Du Z, White RT, Zhang G (2020) Understanding Li-ion cell internal short circuit and thermal runaway through small, slow and *in situ* sensing nail penetration. J Electrochem Soc 167(9):090526

57. Galvez-Aranda DE, Verma A, Hankins K, Seminario JM, Mukherjee PP, Balbuena PB (2019) Chemical and mechanical degradation and mitigation strategies for Si anodes. J Power Sources 419:208–218

58. Jung R, Morasch R, Karayaylali P, Phillips K, Maglia F, Stinner C, Shao-Horn Y, Gasteiger HA (2018) Effect of ambient storage on the degradation of Ni-rich positive electrode materials (NMC811) for Li-ion batteries. J Electrochem Soc 165(2):A132–A141

59. Li Z, Zhang J, Wu B, Huang J, Nie Z, Sun Y, An F, Wu N (2013) Examining temporal and spatial variations of internal temperature in large-format laminated battery with embedded thermocouples. J Power Sources 241(0):536–553

60. Huang S, Wu X, Cavalheiro GM, Du X, Liu B, Du Z, Zhang G (2019) *In situ* measurement of lithium-ion cell internal temperatures during extreme fast charging. J Electrochem Soc 166(14):A3254–A3259

61. Huang S, Du Z, Zhou Q, Snyder K, Liu S, Zhang G (2021) *In situ* measurement of temperature distributions in a Li-ion cell during internal short circuit and thermal runaway. J Electrochem Soc 168(9):090510

62. Zhou H, An K, Allu S, Pannala S, Li J, Bilheux HZ, Martha SK, Nanda J (2016) Probing multiscale transport and inhomogeneity in a lithium-ion pouch cell using *in situ* neutron methods. ACS Energy Lett 1(5):981–986

63. Finegan DP, Scheel M, Robinson JB, Tjaden B, Di Michiel M, Hinds G, Brett DJL, Shearing PR (2016) Investigating lithium-ion battery materials during overcharge-induced thermal runaway: an operando and multi-scale X-ray CT study. Phys Chem Chem Phys 18(45):30912–30919

64. Yokoshima T, Mukoyama D, Maeda F, Osaka T, Takazawa K, Egusa S (2019) Operando analysis of thermal runaway in lithium ion battery during nail-penetration test using an X-ray inspection system. J Electrochem Soc 166(6):A1243–A1250

Chapter 3
Mesoscale Modeling and Analysis in Electrochemical Energy Systems

Venkatesh Kabra, Navneet Goswami, Bairav S. Vishnugopi, and Partha P. Mukherjee

Abstract Electrochemical energy systems are critical from an environmental perspective and provide a pathway to a sustainable energy future. The widespread adoption of these systems is achieved through various applications such as electrically powered aircraft, vehicles, and grid-scale storage. Within these devices, electrochemical physics originates from reaction-coupled interfacial and transport interactions. Advanced computational modeling strategies consider these interactions at multiple temporal and length scales from atomistic to system level. In this context, mesoscale modeling plays a pivotal role in resolving the intermediate length scales, at the intersection of material characteristics and device operation scale. These modeling strategies are contingent upon resolving the fundamental reactive-transport interactions through solving conservation laws. In this chapter, we focus on such a mesoscale modeling methodology accomplished in the context of intercalation electrodes such as lithium-ion batteries, conversion electrodes such as lithium-sulfur batteries, and flow electrodes such as polymer electrolyte fuel cells. The physics-based mass and charge conservation equations are elucidated first which is followed by key examples pertaining to the performance and durability of such systems.

3.1 Introduction

The current level of carbon emissions due to fossil fuel-driven energy production is tipping the scale of global warming toward record temperatures and to the point of no return. Modifying the global landscape of sustainable energy map relies on systematic infrastructure development for energy production and storage for achieving the decarbonization of the environment. Attractive candidates for

V. Kabra · N. Goswami · B. S. Vishnugopi · P. P. Mukherjee (✉)
School of Mechanical Engineering, Purdue University, West Lafayette, IN, USA
e-mail: pmukherjee@purdue.edu

This is a U.S. government work and not under copyright protection in the U.S.;   69
foreign copyright protection may apply 2023
S. Santhanagopalan (ed.), *Computer Aided Engineering of Batteries*,
Modern Aspects of Electrochemistry 62,
https://doi.org/10.1007/978-3-031-17607-4_3

energy production include electrochemical energy systems such as fuel cells and intermittent renewable energy sources such as wind, solar, etc., whereas the storage can be provided by batteries. These advanced electrochemical systems can be deployed to further the sustainability in a variety of objectives, by applying the devices to a gamut of applications including electrically powered aircraft, vehicles, grid storage, etc.

Certain key attributes in the form of suitable energy density, cost, safety, and grid compatibility are necessary to make electrochemical systems such as batteries and fuel cells amenable for large-scale commercialization. Energy storage technology like lithium-ion batteries retain the best combination of the aforementioned characteristics and hence are being adopted in electric mobility applications. In addition, lithium-sulfur batteries typically enhance the range and therefore hold promise in succeeding in the EV market sector as well. Further, hydrogen fuel cells are envisaged for use in heavy-duty transportation owing to aspects such as proper grid scaling and fast refueling capabilities. Thus, there is added impetus to fully transit to a blend of energy storage and hydrogen technology-based sector in order to achieve a clean, decarbonized infrastructure.

The fundamental principle that is ubiquitously prevalent among the various available technologies such as batteries and fuel cells pertains to the electrochemical interactions at the interfacial scale. However, the upscaling of these interactions yields a distinct signature in each case in the form of energy density, power density, and safety window among others. It is noteworthy to mention the tremendous efforts that have been invested in advancing the frontiers of such reliable and environmentally benign technology. The research, in general, has been probed through the synergy of both experimental and computational methods. Of late, this has been aided through data-driven techniques that have facilitated combinatorial studies in addition to assisting material discovery.

Electrochemical storage and conversion devices are inherently multiscale, multiphase, and multiphysics in nature; where the concomitant physics takes place at the interface of the electrode and the electrolyte. Mesoscale modeling is used to capture such critical phenomena that occur at length scales ranging from microns to millimeters. Mesoscale physics thus encompasses the spatial and temporal scales between the material interactions at the nanoscale to the continuum level operation of the electrochemical devices. Mesoscale lies at the intersection of computational tools spanning between two extremes – one being the first principle-based atomistic modeling and the other being macroscale modeling where an apparent tradeoff exists between physics and data. The hierarchy of scales can be bridged often through accurate mesoscale modeling which is akin to a better understanding of the complex behavior of the heterogeneous porous reactors.

In this chapter, we present a discourse on mesoscale modeling of electrochemical storage and conversion devices. The chapter first sheds light on the participating physical phenomena and the generalized transport equations adhered to by the electrochemical systems. The focus on mesoscale modeling with respect to exemplar electrochemical systems like lithium-ion batteries, polymer electrolyte fuel cells, and lithium-sulfur batteries and a few representative highlights have been elaborated in the subsequent sections.

3.2 Electrochemical Physics

3.2.1 Thermo-electrochemical Coupling in Lithium-Ion Batteries

Owing to their high energy density and long life, lithium-ion batteries (LIBs) have garnered tremendous success for portable electronics applications and electric vehicles [1–6]. Over the past years, the focus of Li-ion battery research has been on increasing the energy density and reducing prices to make electric vehicles competitive with gasoline cars. The goals for next-generation batteries are to achieve higher gravimetric energy density, specific energy density, cyclability, and safety. The current research is also looking at beyond Li-ion chemistry including conversion electrode chemistry and solid-state batteries. In this work, we highlight the underlying mesoscale physics, stochastics, thermal-electrochemical-degradation aspects, and safety of energy storage and conversion devices.

A battery involves three structural components, i.e., anode, separator, and cathode, and a working fluid electrolyte. Figure 3.1a shows the working schematic of a Li-ion battery in a pouch cell format. Figure 3.1b represents a magnified view of the electrodes, showing the porous composite electrode and separator. The three materials have similar characteristics and some differences based on their application.

1. Anode/Cathode: The role of the electrode is to store the lithium ions and facilitate an active reaction surface for electrochemical reactions to occur. The generated Li^+ ions then move through tortuous porous pathways of the electrode to another electrode. Meanwhile, the generated electrons are conducted through the solid phase toward the current collector. The microstructural view of the electrode is shown in Fig. 3.1b. A quick observation suggests that the particle morphology of the electrode is very different across the two electrodes. The particle size for anode typically ranges between 5 and 15 micrometers while for cathode between 2 and 10 micrometers. These particle shapes and sizes affect the performance of the battery and can be tailored for different applications.
2. Separator: It is made up of porous polyolefin films, nylon, or cellophane. The separator has a fibrous structure with pores ranging between 30 nm and 100 nm as shown in Fig. 3.1b. The separator has a very low electronic conductivity which stops any flow of electronic current between the electrodes. It enables the passage of Li^+ ions through the pores, shuttling between the electrode during charging and discharging operations.

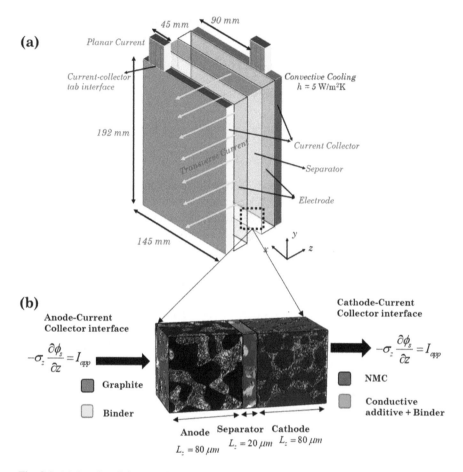

Fig. 3.1 (**a**) Pouch cell format of a lithium-ion battery. (Reproduced with permission from Ref. [7]. Copyright 2021 Elsevier), (**b**) Schematic representation of the NMC333-LiC6 graphite Li-ion full cell sandwich model. (Reproduced with permission from ref. 8. Copyright 2020 American Chemical Society)

The electrodes in a battery are not limited to a single material. Figure 3.1b shows that different materials are added to the battery electrodes which enable efficient functioning of the electrodes [9–12]. Based on the role of electrodes, we see that they need to provide a high surface area for the reaction which is achieved by using porous electrodes. The conductive additives are added to the cathodes due to their poor intrinsic electronic conductivity of the cathode, enabling a minimal resistance path to electrons generated at the electrode needing to be conducted through the solid phase to the current collector. In addition, binders are added to both the electrodes to enable uniform film formation and keep the electrode particles bonded together. Figure 3.1b pictorially represents these multiple phases present in an electrode.

The four major steps in Li-ion battery operation such as charging/discharging are:

1. Diffusion: The diffusion of Li within the active material particles of the electrodes.
2. Kinetics: Deintercalation/intercalation reaction at the interface of electrode and electrolyte.
3. Transport of Li^+ ions: The transport of Li^+ ions through the separator and electrode pores.
4. Conduction of electrons: The passage of electrons through the solid electrode phase.

Considering the active material particles to be spherical, which is true to a good extent and helps in reducing the computational complexity of the model, these active particles act as a storage for Li to be available for reaction. We model the diffusion of Li within the particles using Fick's law [13], that is:

$$\frac{\partial C_s}{\partial t} = \frac{1}{r^2}\frac{\partial}{\partial r}\left(D_s r^2 \frac{\partial C_s}{\partial r}\right) \tag{3.1}$$

$$BC : @r = 0 : \frac{\partial C_s}{\partial r} = 0, \quad @r = R_s : \frac{\partial C_s}{\partial r} = \frac{-i}{D_s F} \tag{3.2}$$

We solve for the radial concentration field within the active particles. Here subscript "s" denotes the fields in the solid phase. "C_s" denotes the concentration of Li within the active material particles, and "D_s" represents the diffusivity of Li within the active particles. The boundary conditions at $r = 0$ are the results of the symmetry of the reaction current density at the core of the particle. The boundary condition at $r = R$ is the balance of incoming/outgoing flux of Li due to the electrochemical redox reaction at the surface of the active particle.

Our system of interest primarily undergoes intercalation reaction or electrochemical redox reaction at the surface of the active particle. The reaction can be represented as:

$$Li_{1-x}C_6 + xLi^+ + xe^- \underset{discharge}{\overset{charge}{\rightleftharpoons}} Li_x C_6 \tag{3.3}$$

The rate of intercalation reaction is modeled using the Butler-Volmer equation and its expression is given by [13]:

$$i = k\sqrt{C_s C_e \left(C_s^{max} - C_s\right)}\left\{e^{F\eta/2RT} - e^{-F\eta/2RT}\right\} \tag{3.4}$$

where C_s^{max} is the maximum Li concentration (mol/m^3) which can be stored in the active particles and C_e is Li^+ concentration within the electrolyte phase. "k"

is the temperature-dependent intercalation reaction constant of the reaction. The overpotential $\eta = \phi_s - \phi_e - U$ is the driving force for intercalation/deintercalation reaction, ϕ_s represents the solid-phase potential, and ϕ_e represents the electrolyte-phase potential with $U(C_s)$ being open circuit potential of the electrode material. "i" represents the areal reaction current density in the electrode, and a positive value represents the generation of Li$^+$ ions, while the negative value represents the consumption of Li.

The species conservation in the electrolyte phase is utilized for solving the (C_e) electrolyte concentration variation within the electrode [13]. It is modeled using the Nernst-Planck equation:

$$\varepsilon \frac{\partial C_e}{\partial t} = \nabla \cdot \left(D_e \frac{\varepsilon}{\tau} \nabla C_e \right) + \frac{(1 - t_+) a_s j}{F} \tag{3.5}$$

Here, "t_+" is the electrolyte transference number that describes the part of current transported by the lithium ions and usually a constant. Also, "a_s" represents the interfacial area of the electrode per unit volume.

The charge conservation in the electrolyte phase is given by [13]:

$$\nabla \cdot \left(K_e \frac{\varepsilon}{\tau} \nabla \phi_e \right) + \nabla \cdot \left(K_d \frac{\varepsilon}{\tau} \nabla \ln C_e \right) + a_s j = 0 \tag{3.6}$$

The Li$^+$ ions flow results in the ionic current in the electrode pores. This equation describes the appropriate constitutive relations for the ionic flux in terms of migration and diffusion current, with the absence of electrolyte advection current. The diffusion components depend on the concentration gradients of Li$^+$ ions, and migration depends on the electric potential gradients in the electrolyte. The term "K_e" stands for ionic conductivity, and "K_d" stands for diffusional conductivity.

Charge conservation within the solid matrix follows Ohm's law as well [13]:

$$\nabla \cdot \left(\sigma_s^{\text{eff}} \nabla \phi_s \right) - a_s j = 0 \tag{3.7}$$

Here the σ_s^{eff} represents the effective conductivity of the solid skeleton and includes all solid phases.

Each of these processes is represented by a partial differential equation and is closely coupled to each other. To obtain a unique and well-defined solution, appropriate boundary conditions and continuity conditions are ensured at the domain interfaces. The boundary condition is expressed in terms of the applied current density, which is expressed in terms of the C-rate. Theoretically, C-rate is the inverse of the time of operation of the battery, i.e., C/2 implies the battery will be charged/discharged in 2 h. C-rate is relative to a battery's electrode capacity. The electrode capacity is the amount of charge that can be stored in the porous electrode, defined by the following relation:

$$Q = \frac{F C_s^{max} \varepsilon_s L}{3600} \left[A/m^2 \text{ of electrode cross-section} \right] \tag{3.8}$$

and equivalently current density can be expressed as:

$$I_{app} = \text{C-rate} \cdot Q \tag{3.9}$$

Boundary conditions at the electrode-current collector interface:

$$-D_e \frac{\varepsilon}{\tau} \nabla C_e = 0 \tag{3.10}$$

$$\frac{\partial \phi_e}{\partial x} = 0 \tag{3.11}$$

$$-\sigma_s^{eff} \frac{\partial \phi_s}{\partial x} = I_{app} \tag{3.12}$$

In a battery, the porous electrode is sandwiched between the separator and the current collector. So, the electrode acts as a reactor converting the purely electronic current from the current collector into purely ionic at the separator. At any location in the electrode, while the total current which is the sum of ionic and electronic current is fixed, the reaction current distribution can exhibit spatial variation. At the electrode-separator interface, the current is completely ionic, and thus current flux continuity needs to be ensured as given below:

$$\left(-D_e \frac{\varepsilon}{\tau} \nabla C_e \right)_{x=L_e-\delta} = \left(-D_e \frac{\varepsilon}{\tau} \nabla C_e \right)_{x=L_e+\delta} \tag{3.13}$$

$$\left(\kappa \frac{\varepsilon}{\tau} \nabla \phi_e + \kappa_D \frac{\varepsilon}{\tau} \nabla \ln C_e \right)_{x=L_e-\delta} = \left(\kappa \frac{\varepsilon}{\tau} \nabla \phi_e + \kappa_D \frac{\varepsilon}{\tau} \nabla \ln C_e \right)_{x=L_e+\delta} \tag{3.14}$$

$$\left(\frac{\partial \phi_s}{\partial x} \right)_{L_e-\delta} = \left(\frac{\partial \phi_s}{\partial x} \right)_{L_e+\delta} = 0 \tag{3.15}$$

Overall, the electrochemical dynamics of an intercalation electrode is governed by the set of coupled partial differential Eqs. (3.1), (3.5), (3.6), and (3.7) with reaction kinetics from Eq. (3.4) and boundary conditions from Eqs. (3.10), (3.11), and (3.12). The composite electrode consists of multiple phases with different length scales. The impact of these multiscale and multiphase characteristics of the electrode is accounted for through effective properties such as active area, a; porosity, ε; tortuosity, τ; and effective conductivity, σ^{eff}. The values of these are obtained from the electrode characterization as would be discussed later in Sect. 3.3.1.1.

Within a porous electrode, the current is carried by multiple charge carriers through different modes of transport, i.e., electronic current by electrons, ionic current by Li^+ ions, and the interconversion between these carriers through an electrochemical reaction. Each carrier experiences resistance to its motion and equivalently leads to irreversibility and joule heating.

Based on the various physical phenomena occurring within the battery, the heat generation is segregated into different modes:

Ohmic heat:

$$\dot{q}_{ohmic} = \left(\kappa^{eff} \nabla \phi_e \cdot \nabla \phi_e + \kappa_D^{eff} \nabla \ln C \cdot \nabla \phi_e \right) + \sigma^{eff} \nabla \phi_s \cdot \nabla \phi_s \qquad (3.16)$$

Kinetic heat:

$$\dot{q}_{kinetic} = a_s j \left(\phi_s - \phi_e - U \right) \qquad (3.17)$$

The ohmic heat (Eq. 3.16) is the result of resistance due to transport in bulk phases, i.e., it is the sum of heat generation in electrolyte and solid phases, while the kinetic heat (Eq. 3.17) relates to interfacial reactions. Both are irreversible in nature and jointly result in total Joule heating. The third mode of heat generation which is reversible in nature arises due to the entropy of electrochemical reaction.

Entropic heat:

$$\dot{q}_{entropic} = -a_s j T \frac{dU}{dT} \qquad (3.18)$$

The entropic heat (Eq. 3.18) is generated at a very low rate, while the other heat generation approaches zero. The heat generation at the electrode is a strong function of the pore-scale interaction and material properties; thus, microstructural attributes and operating conditions play an important role in modulating the overall thermal response [14].

3.2.2 Physicochemical Interactions in Lithium-Sulfur Battery Electrodes

Lithium-sulfur (Li-S) chemistry offers tremendous promise for energy storage applications due to their high theoretical capacity (1675 mAhg^{-1}), low material cost, and the earthly abundance of sulfur [40–44]. In contrast to conventional lithium-ion batteries, Li-S batteries involve an electrolyte-based reaction paradigm that bypasses the sluggish kinetics of solid-state intercalation. During electrochemical operation, Li-S battery electrodes undergo a series of microstructural changes based on the precipitation/dissolution interactions that take place in the reaction pathway

[45, 46]. Based on variations in electrode composition and precipitate morphology, pore-scale characteristics such as electrochemically active area and ion percolation pathways are quantitatively different. Such changes in the pore network can alter mechanisms like species transport evolution, pore-phase blockage, and surface passivation. The reaction pathway in the Li-S battery typically consists of the following set of chemical and electrochemical reactions. Herein, the forward and backward directions represent the reactions taking place during the discharge and charge of the cell, respectively.

$$\text{Chemical}: \quad S_{8(s)} \rightleftharpoons S_{8(l)} \tag{3.19}$$

$$\text{Electrochemical}: \quad \frac{1}{2}S_{8(l)} + e^- \rightleftharpoons \frac{1}{2}S_8^{2-} \tag{3.20}$$

$$\text{Electrochemical}: \quad \frac{1}{2}S_8^{2-} + e^- \rightleftharpoons S_4^{2-} \tag{3.21}$$

$$\text{Electrochemical}: \frac{4}{3}Li^+ + \frac{1}{6}S_4^{2-} + e^- \rightleftharpoons \frac{2}{3}Li_2S_{(s)} \tag{3.22}$$

Here, Eq. (3.19) represents the chemical dissolution of the solid sulfur, which is dependent on its solubility in the electrolyte. The sequential reduction of the dissolved sulfur to higher-chain polysulfide and higher to medium-chain polysulfide is represented by Eq. (3.20) and Eq. (3.21), respectively. Eq. (3.22) represents the formation of Li$_2$S precipitate through the reduction of the medium-chain polysulfide. Based on the described reaction pathway, the species balance for the four individual species, namely, dissolved sulfur, Li$^+$, long-chain polysulfide, and medium-chain polysulfide, can be mathematically represented as follows:

$$\frac{\partial (\varepsilon C_i)}{\partial t} = -\nabla.N_i + \dot{R}_i \tag{3.23}$$

Herein, ε is local porosity, N_i is species flux, C_i is species concentration, and $\dot{R}_i$ is local source rate of the species, all of which dynamically evolve upon cell operation.

Governing equations for ionic current and electronic current conservation are expressed in Eq. (3.24) and Eq. (3.25), respectively:

$$\nabla.\left(k\nabla\phi_e + k_{Li^+}\nabla \ln C_{Li^+} + k_{S_8^{2-}}\nabla \ln C_{S_8^{2-}} + k_{S_4^{2-}}\nabla \ln C_{S_4^{2-}} + k_{S_8(l)}\nabla \ln C_{S_8(l)}\right)$$
$$+ \dot{j} = 0$$
$$\tag{3.24}$$

$$\nabla.\left(\sigma^{\text{eff}}\nabla\phi_s\right) - \dot{j} = 0 \tag{3.25}$$

ϕ_e and ϕ_s correspond to the electrolyte and solid-phase potential, respectively, and $\dot{J}$ represents the volumetric reaction current density, which is equal to the sum of the reaction current density for the electrochemical reactions in Eqs. (3.26), (3.27), and (3.28). Conservation of the solid phases ($Li_2S_{(s)}$ and $S_{8(s)}$) and the electrolyte pore space is governed by the following expressions:

$$\frac{\partial \left(\varepsilon_{S_8}\right)}{\partial t} = -\tilde{V}_{S_8} \dot{R}_{S_8\uparrow} \tag{3.26}$$

$$\frac{\partial \left(\varepsilon_{Li_2S}\right)}{\partial t} = \tilde{V}_{Li_2S} \dot{R}_{S_4^{2-}\rightarrow Li_2S} \tag{3.27}$$

$$\frac{\partial \left(\varepsilon_{S_8} + \varepsilon_{Li_2S} + \varepsilon\right)}{\partial t} = 0 \tag{3.28}$$

ε_{S_8} and ε_{Li_2S} are the volume fractions of the solid phases of sulfur and Li_2S precipitate, which are dependent on their molar volumes ($\tilde{V}_{S_8}$, $\tilde{V}_{Li_2S}$) and rates of consumption/production ($\dot{R}_{S_8\uparrow}$, $\dot{R}_{S_4^{2-}\rightarrow Li_2S}$). As presented in Eq. (3.29), the sum of volume fractions of the sulfur utilized, precipitate formed, and the electrolyte porosity is conserved throughout the cell operation. The applied current density of the system (J_{app}) is defined as follows:

$$J_{app} = C_{rate} \left(\frac{16 F \varepsilon_{S_8}^0 L_{cathode}}{3600 \, \tilde{V}_{S_8}}\right) \tag{3.29}$$

Here, $\varepsilon_{S_8}^0$ and $L_{cathode}$ refer to the pristine porosity and length of the cathode, respectively.

3.2.3 Multiphase, Multicomponent Transport in Polymer Electrolyte Fuel Cells

Polymer electrolyte fuel cells (PEFCs) are being considered as a promising clean energy source to serve various target applications like automobiles as well as stationary and portable energy systems [49]. This can be primarily attributed to the fact that PEFCs are environmentally benign, have low noise operation, and have appreciable energy conversion efficiency. Concerted efforts are being made into the design and engineering space of PEFC technology for enhancing the performance and durability to enable its adoption in light- and heavy-duty vehicles and other commercial applications including residential supply.

As seen in the schematic shown in Fig. 3.2, the working of a PEFC is based on the electrochemical principle of converting hydrogen and oxygen into the water

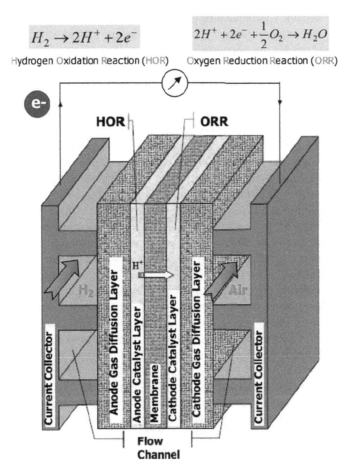

Fig. 3.2 Schematic diagram of a polymer electrolyte fuel cell. (Reproduced with permission from Ref. 50. Copyright 2008 RSC Publishing)

to produce energy. A typical setup consists of a membrane electrode assembly (MEA) wherein a proton-conducting (H^+) membrane made of polymeric material is sandwiched between two electrodes, an anode and a cathode. In addition, individual bipolar plates are present on either face of the MEA device. Both the electrodes (anode and cathode) are further comprised of a catalyst layer (CL) and a gas diffusion layer (GDL) with a microporous layer (MPL) cast at the interface. A flow channel directs gaseous reactants onto the gas diffusion layer/microporous layer (GDL/MPL) and finally into the electrochemically active catalyst layer (CL). The incoming feeds of the anode and cathode generally consist of hydrogen/reformate gas and oxygen/air (humidified) stream, respectively. The functionality of the GDL is to serve two purposes – firstly, to allow the ingress of the reactants (including fuel and oxidant) and the egress of the by-product (liquid water) and, secondly, to provide mechanical rigidity to the electrode. The size of the catalyst layer is of the order of a few microns. Inside its porous structure, Pt-based catalyst nanoparticles are supported on high surface area carbon substrates with a heterogeneous dispersion of ionomer. These nanoparticles are preferred for their strong catalytic activity and high specific surface area with the reactions occurring at the triple-phase boundary of ionomer, Pt, and carbon. Electrochemical reactions occur only inside the CL; thus carbon particle sizes inside the CL are smaller in size as compared to GDL, to enhance the surface area.

Hydrogen oxidation reaction (HOR) occurring at the anode catalyst layer (ACL) exhibits facile reaction kinetics, whereas oxygen reduction reaction (ORR) occurring at the cathode catalyst layer (CCL) is reported to be one of the slowest reactions in electrochemistry. The sluggishness associated with the ORR poses a dominant kinetic limitation of the PEFC and leads to considerable voltage loss. The robust performance of the PEFC is also predicated on appropriate water management. Low humidification/water content results in poor proton conductivity in the membrane thereby emanating ohmic losses. On the other hand, excess water content leads to flooding of the PEFC. This indicates passivation of the reaction sites in the CCL in addition to hindering the transport of oxygen through the pores resulting in a net higher concentration overpotential or mass transport limitations. Thus, performance enhancement of a PEFC is possible through rational design changes that connect key aspects of CCL in terms of architecture, performance, and durability.

A PEFC model is built on the conservation laws for mass, momentum, species, charge, and thermal energy for both single-phase and two-phase transport. These coupled equations are numerically solved in order to obtain the flow-field, species concentrations, potentials, and the liquid water dynamics throughout an operating fuel cell. The governing equations enumerated below are motivated by the M^2 model for flow and transport in porous media [51–55].

Based on M^2 model, the mixture properties need to be derived which are all dependent on the liquid saturation defined as the ratio of the liquid volume to the pore volume (Eq. 3.30). Accordingly, the mixture density, ρ, and molar concentration, C_i, can be obtained from Eqs. 3.31 and 3.32:

$$s = \frac{V_l}{V_{pore}} \tag{3.30}$$

$$\rho = \rho_l s + \rho_g (1 - s) \tag{3.31}$$

$$C_i = C_{i,l} s + C_{i,g} (1 - s) \tag{3.32}$$

Conservation of Mass Mass conservation for the mixture is given by:

$$\nabla \cdot (\rho \mathbf{u}) = 0 \tag{3.33}$$

Here, ε is the porosity of the medium, and u is the superficial mixture velocity. Under a single-phase approximation, u would correspond to gaseous mixture velocity.

Conservation of Momentum Momentum conservation for the mixture based on mixture velocity u is:

$$\frac{1}{\varepsilon^2} \nabla \cdot (\rho uu) = \nabla \cdot (\mu \nabla u) - \nabla p - \frac{\mu}{K} u \tag{3.34}$$

Here, K is the absolute permeability of the porous medium, and mixture viscosity μ is defined as:

$$\mu = \rho \left[\frac{k_{rl}}{\nu_l} + \frac{k_{rg}}{\mu_g} \right] \tag{3.35}$$

where k_{rl}, k_{rg}, and ν_l, ν_g are the relative permeabilities and kinematic viscosities of liquid and gas-phase, respectively. Relative permeabilities of individual phases are related to the cube of phase saturations such that:

$$k_{rk} = s_k^3 \tag{3.36}$$

$$k_{rg} = (1-s)^3 \tag{3.37}$$

where $(1-s)$ corresponds to the pore occupation fraction of the gas phase.

The source term in the momentum equation is the Brinkman extension to Darcy's law for porous media using superficial velocities which obey interface continuity.

Conservation of Species Species conservation equation is defined in terms of molar concentration such that:

$$\nabla \cdot (\gamma_c uC_i) = \nabla \cdot \left(D_{i,g}^{eff} \nabla C_{i,g} \right) - \nabla \cdot \left[\left(\frac{C_{i,l}}{\rho_l} - \frac{C_{i,g}}{\rho_g} \right) j_l \right] + S_i \tag{3.38}$$

Here, C_i is the total concentration of species i in liquid and gas phases for two-phase flows, and D_i is the diffusivity of species i. The superscript *eff* denotes the effective diffusivity for the macrohomogeneous form of the porous media. Fick's law has been utilized for describing multicomponent diffusion which is exact for binary diffusion and is a good approximation for multicomponent diffusion.

Due to the different flow fields of the liquid and gas phase, species advection needs to be corrected by the advection factor γ_c defined as:

$$\gamma_c = \begin{cases} \dfrac{\rho}{C_{H_2O}}\left(\dfrac{\lambda_l}{M_{H_2O}} + \lambda_g \dfrac{C_g^{H_2O}}{\rho_g}\right) & \text{for water} \\[2ex] \dfrac{\rho\lambda_g}{\rho_g(1-s)} & \text{for other species} \end{cases} \tag{3.39}$$

where λ_l, λ_g are the relative mobility of liquid and gas phases, respectively:

$$\lambda_l = \frac{k_{rl}/\nu_l}{k_{rl}/\nu_l + k_{rg}/\nu_g} \tag{3.40}$$

$$\lambda_g = 1 - \lambda_l \tag{3.41}$$

The second term on the right-hand side in the species conservation equation accounts for capillary transport due to surface tension effects in porous media with the capillary flux j_l defined as:

$$j_l = \frac{\lambda_l \lambda_g}{\nu} K \nabla p_c \tag{3.42}$$

Here p_c is the capillary pressure defined as the difference in pressure between the gas phase and the liquid phase and is given by:

$$p_c = \sigma \cos(\theta_c) \left(\frac{\varepsilon}{K}\right)^{0.5} J(s) \tag{3.43}$$

with σ denoting surface tension, θ_c the equilibrium contact angle, and $J(s)$ being the Leverett function for both hydrophobic and hydrophilic GDL dependent on liquid phase saturation and contact angle as highlighted below:

$$J(s) = \begin{cases} 1.417\,(1-s) - 2.120(1-s)^2 + 1.263(1-s)^3 & \text{if } \theta_c < 90° \\ 1.417s - 2.120s^2 + 1.263s^3 & \text{if } \theta_c > 90° \end{cases} \tag{3.44}$$

The final term on the right-hand side of the species conservation equation, S_i, is the source/sink term of species i from its generation/depletion by electrochemical reactions and electroosmotic drag contributions. Electroosmotic drag of water arises from the motion of the water alongside H^+ due to the formation of a solvated complex of hydronium ions, H_3O^+. Consequently, the source term due to electrochemical reaction and electroosmotic drag can be written as:

$$S_i = -\frac{s_i\, j}{nF} - \nabla\left[\frac{n_d}{F} i_e\right] \tag{3.45}$$

The first term in the above equation signifies the depletion term due to the electrochemical reaction current density j with the electrochemical reaction being represented as $\sum_i s_i M_i^z = ne^-$, where M_i is the chemical formula of species i, s_i is the stoichiometric coefficient, and n is the number of electrons transferred. This term exists only in the catalyst layer and is zero for the flow channels, gas diffusion layer, and ionomeric membrane. The last term accounts for the electroosmotic drag of water alongside protons due to ionic current in the electrolyte i_e with the electroosmotic drag coefficient n_d. It should be noted that n_d is only relevant for the water molecule and its value is set to zero for other transported species like H_2 and O_2. Electroosmotic drag term is non-zero only where ionic current exists, i.e., in the catalyst layers and ionomeric membrane, and is zero elsewhere.

For modeling of single-phase flows, the effect of advection factor and capillary transport does not exist, and hence the species conservation equation can be reduced to the following form:

$$\nabla \cdot (uC_i) = \nabla \cdot \left(D_{i,g}^{\text{eff}} \nabla C_{i,g}\right) + S_i \tag{3.46}$$

Conservation of Charge Charge conservation in PEFCs is described by electrolyte and solid-phase potential equations in the ionomer phase and Pt-C phase, respectively:

$$\nabla \cdot \left(\kappa^{\text{eff}} \nabla \phi_e\right) + j = 0 \tag{3.47}$$

$$\nabla \cdot \left(\sigma^{\text{eff}} \nabla \phi_s\right) - j = 0 \tag{3.48}$$

Here, κ and σ are the ionic and electronic conductivity of the electrolyte and solid-phase, respectively. The membrane and catalyst layer ionic conductivity is generally a strong function of the water content. For large electronic conductivity of the solid phase, the anode and cathode solid-phase potentials can be assumed to be constant and given by $\phi_{s,a} = 0$ and $\phi_{s,c} = V_{\text{cell}}$.

Electrochemical current density at the electrode-electrolyte interface is described by Butler-Volmer kinetics for a general electrochemical reaction. For PEFCs with rapid hydrogen oxidation reaction kinetics (HOR) at the anode and sluggish oxygen reduction reaction (ORR) at the cathode, the Butler-Volmer equation reduces to the linear regime at the anode and Tafel regime in the cathode, respectively, given by:

$$j_a = a j_{0,a}^{\text{ref}} \left(\frac{c_{H_2}}{c_{H_2,\text{ref}}}\right)^{1/2} \left(\frac{\alpha_a + \alpha_c}{RT} F \eta\right) \tag{3.49}$$

$$j_c = a j_{0,c}^{\text{ref}} \left(\frac{c_{O_2}}{c_{O_2,\text{ref}}}\right) \exp\left(-\frac{\alpha_c}{RT} F \eta\right) \tag{3.50}$$

Here, a is the specific active surface area of the catalyst layer of the electrodes, j_0 is the exchange current density, and α_a, α_c are the anodic and cathodic transfer coefficients for each electrochemical reaction with $\alpha_a + \alpha_c = 2$ for the HOR and $\alpha_c = 1$ for ORR. Overpotential η depends on the solid-phase potential and electrolyte-phase potential along with the open circuit potential (OCP) for each electrode given by $\eta_a = \phi_s - \phi_e$ and $\eta_c = \phi_s - \phi_e - U_{oc}$ for anode and cathode, respectively. Here, we have utilized the fact that the OCP of the standard hydrogen electrode is equal to zero. OCP of the cathode is a function of temperature and partial pressure given by:

$$U_{oc} = 1.23 - 0.9 \times 10^{-3}\,(T - 298) + \frac{RT}{F}\left(\ln p_{H_2} + \frac{1}{2}\ln p_{O_2}\right) \tag{3.51}$$

Conservation of Energy The energy conservation for two-phase flow in PEFCs is given by:

$$\frac{\partial\left[(\rho c_p)_m T\right]}{\partial t} + \nabla \cdot \left[\gamma_h (\rho c_p)_f \mathbf{u} T\right] = \nabla \cdot \left(k^{eff} \nabla T\right) + S_T \tag{3.52}$$

Here, $(\rho c_p)_m$ represents the total heat capacitance of the porous medium (m) considering both solid matrix(s) and two-phase mixture in the pores (f) given by $(\rho c_p)_m = \varepsilon (\rho c_p)_f + (1 - \varepsilon)(\rho c_p)_s$ Fourier's law of heat conduction accounts for conductive heat transfer with k^{eff} being the effective thermal conductivity of the porous medium. The last term on the right-hand side represents the heat source term arising from heat generation due to kinetic heat of the electrochemical reaction $(j\eta)$, reversible (or entropic) heat $\left(jT\frac{dU}{dT}\right)$, and ohmic heat from the transport of electronic current (i_s^2/σ^{eff}) and ionic current (i_e^2/κ^{eff}) . Additional heat generation/depletion can arise from condensation and evaporation of water vapor into liquid water and vice versa. Consequently, the heat generation source term takes the form:

$$S_i = j\eta + jT\frac{dU}{dT} + \frac{i_s^2}{\sigma^{eff}} + \frac{i_e^2}{\kappa^{eff}} + \dot{m}_{fg}h_{fg} \tag{3.53}$$

3.3 Mesoscale Modeling with Case Studies in Exemplar Electrochemical Systems

3.3.1 Lithium-Ion Batteries

3.3.1.1 Porous Microstructure Characterization: Effective Property Calculations

The porous and multiphase nature of the battery electrode (see Fig. 3.3a) leads to a significantly complicated structure from a modeling standpoint. The performance of porous electrodes can be characterized in terms of their effective properties,

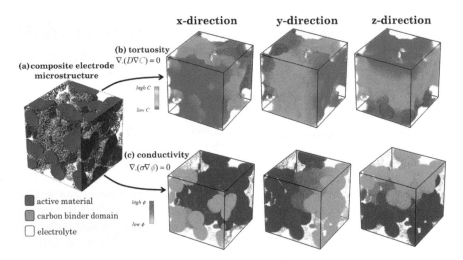

Fig. 3.3 Porous media characterization for (**a**) composite electrode microstructures, (**b**) pore network tortuosity, and (**c**) solid-phase effective electronic conductivity in all three orthogonal directions. Due to a smaller dimension of carbon and binder phases, it is treated as a pseudo phase – and referred to as carbon binder domain (CBD) [15]. The material conductivity for this pseudo phase is a function of carbon-to-binder ratio. AB = acetylene black conductive additives and PVDF = polyvinylidene fluoride binder [15]. (Figure reproduced with permission from Ref [15] Copyright (2018) (American Chemical Society))

namely, – a, ε, τ and σ_{eff}. The interfacial area (a) and porosity (ε) calculation of the electrode is relatively straightforward [15]. The calculation of the effective conductivity and tortuosity is involved, and discussed below. The tortuosity is used to quantify the transport efficiency of Li$^+$ ions through the pores; meanwhile, conductivity measures the connectivity of solid phase to facilitate electron conduction. Direct numerical simulations (DNS) are used to compute these properties.

Figure 3.3b shows that for solving the tortuosity, we need to solve the concentration balance equation in the composite electrode. Pores are assigned suitable diffusivity values, while the solid phase has zero diffusivity value for their transport contributions. Principally identical conductivity calculation (see Fig. 3.3c) can be performed with solid-phase assigned suitable conductivity and the pore phase having zero conductivity, as we know that composite electrode involves two different solids – conductive additives and binders. The length scale of these conductive additives and binder is much smaller than the active material particle sizes. Due to this, they can be assumed to be jointly forming a pseudo phase. So, properties of interest such as density and conductivity have now become a function of conductive additive to binder ratio [15]. The secondary phase's spatial arrangement also plays a critical role in determining the electrode properties. The spatial arrangement of the secondary phases can range from finger-like to film-like morphology. The arrangement of these secondary phases affects the available interfacial area and pore network arrangement. The finger-like arrangement exposes a higher interfacial area at the electrode but causes more hindrance (due to finger-like projections) to pore space. These arrangements can be optimized to improve the performance of the Li-ion battery without altering the electrode materials [15].

3.3.1.2 Thermo-electrochemical Interactions

The electrode microstructure at the anode and cathode are very different in a full cell, along with the material differences that complicate the analysis. A typical anode is made up of graphitic platelet-shaped particles, and cathode NMC has spherical particles. Figure 3.4a,b denotes the typical differences in electrode microstructure for the anode and cathode. Differences in particle morphology significantly impact the pore network and notable impact on the kinetic and transport property of electrodes. Figure 3.4c,d shows this is, in turn, affecting the heat generation and electrochemical performance characteristics. We observe that the heat generation is quite asymmetric between anode and cathode. The cathode has a high kinetic heat generation because the spherical particles offer the least surface area for a given volume. On other hand, the anode offers higher ohmic heat generation due to the platelet-shaped particles significantly distorting the pore space and leading to high pore phase transport resistance. Figure 3.4e,f shows the difference in capacity and temperature rise during the charging and discharging of a Li-ion battery. Temperature rise during charging shows a quadratic trend with the C-rate; meanwhile temperature rise during discharging is increasing with saturation at 5C, suggesting quadratic behavior at higher C-rates. Temperature rise is a function of heat generation rate and capacity; heat generation rate increases with C-rate; meanwhile, the capacity decreases. In the discharging process, the heat generation rate is higher compared to the charging process, leading to a higher total heat generation and, thus, temperature rise. The increased temperature rise is beneficial in improving the temperature-dependent rate constants and diffusivities within the electrodes. This causes higher capacity during discharging compared to charging. This phenomenon is called thermo-electrochemical hysteresis, due to the thermal effects leading to the different charge and discharge characteristics.

The inherent coupling of the thermal and electrochemical interaction is characteristic of the high-energy cells, i.e., where a high amount of active material is tightly packed. Consider Fig. 3.5a showing the increased active material loading from 40% to 70%. We observe that the cell voltage and temperature increase rapidly as the packing density of the active material is increased up to 60% packing. Beyond this packing limit, cell shutdown takes place even before meaningful capacities are reached. On increasing the packing density, the pores get constricted, and their transport resistance increases significantly; this can be observed from Fig. 3.5b where the heat generation rates show an exponential increase with packing. Since the temperature rise is a function of heat generation rate and the charging capacity, we observe temperature rise is highest for 70% of the packed electrode. Either packing too densely causes abrupt shutdown of the electrode or too sparse packing leads to smaller heat generation rates.

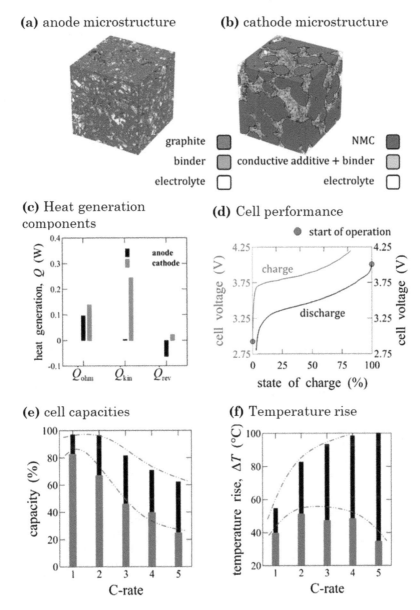

Fig. 3.4 Charge-discharge behavior due to dissimilar microstructures at the anode and cathode, the direction of current flow (i.e., charging or discharging). (**a**) Representative anode microstructure with 95 wt % graphite and (**b**) cathode has 90 wt % NMC, (**c**) heat generation components, (**d**) cell voltage during charging and discharging, (**e**) cell capacity exhibiting different charging-discharging trends with C-rate, (**f**) cell temperature rise exhibiting different charging-discharging trends with C-rate. (Reproduced with permission from ref. 16. Copyright 2018 American Chemical Society)

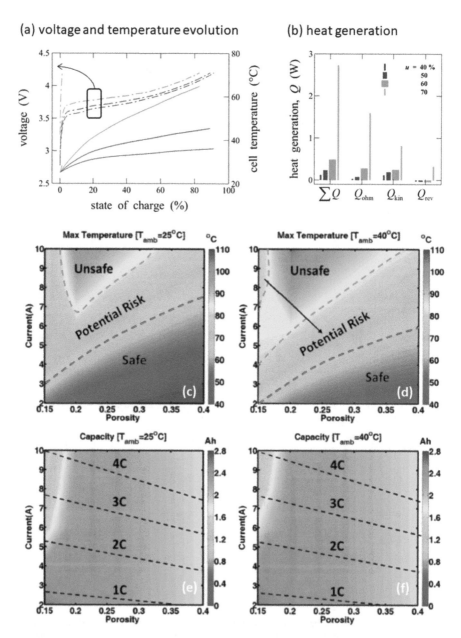

Fig. 3.5 Thermal excursion and safety risk in high energy density cells (**a**) cell voltage, temperature increase due to higher overpotential in highly packed electrodes; (**b**) Heat generation increases monotonically with active material packing (Reproduced with permission from ref. 16. Copyright 2018 American Chemical Society); (**c–f**) Impact of porosity, applied current under different ambient temperature shows varying temperature rise and capacity trends [17]

This intricate coupling of electrode microstructure, operating rate, and ambient temperature has been explored in Fig. 3.5c–f. Based on the different ambient temperatures of the cell, one can identify a safe operating zone based on thermal safety requirements. We observe that a low porosity electrode alternatively highly packed electrode has higher heat generation and smaller capacity, leading to potential risk conditions. The abrupt shutdown is caused by a lack of ions in the electrolyte to sustain electrochemical reactions. Whereas electrodes with moderate packing are at the highest risk of unsafe operation due to significant temperature rise. Electrodes with the least dense packing provide less risk of thermal runaway.

3.3.1.3 Degradation Phenomena in Lithium-Ion Batteries

Lithium-ion battery systems are promising energy conversion and storage technology and have proven to be useful for electric vehicle applications to grid energy storage. The batteries have a limited lifespan because of the degradation of the active material particles in the electrode caused by repeated charging and discharging. The primary modes of degradations are microcrack formation within particles due to diffusion-induced stresses and chemical/electrochemical solid-electrolyte interphase growth [17–22]. These effects and coupling between them lead to capacity fade and impedance growth leading to cell performance and lifetime decline.

Fast charging is also one of the prominent challenges for LIBs to be able to compete with gasoline counterparts in terms of refueling time. The extreme fast charging (XFC) refers to the charging rates including and above 4C (charging time below 15 min), inevitably leading to an accelerated rate of degradation. At higher rates, the side reaction of Li^+ ions directly leads to reduction and deposition onto graphitic anode particles known as Li-plating. Such side reactions can lead to permanent loss of Li inventory and is a leading cause for capacity fade [23–28]. Another challenge for the fast charging of cells is excessive heat generation, which can cause rapid temperature rise leading to thermal runaway.

Li-Plating and Solid Electrolyte Interphase (SEI) Formation

Lithium-ion batteries (LIBs), with intercalation chemistry, store their chemical energy as lithium in the anode and cathode particles. During the charging operation of the battery, the Li diffuses to the surface of the cathode's active material particles, wherein it undergoes oxidation reaction and forms Li^+ ions which migrate and diffuse along with the cathode and separator. Upon reaching the anode, these ions undergo electrochemical reduction to Li followed by intercalation and diffusion into the anode particle. Within the extreme fast-charging regime, the slow diffusion ($\sim 10^{-14}$ [m^2/s]) and sluggish kinetics cause excess of Li^+ ions to reduce over the anode resulting in Li-plating. Over multiple charging cycles, the Li-plating layer grows and causes irreversible loss of Li, causing capacity fade and impedance rise. Several experiments have established the presence of Li-plating through characterization techniques based on nuclear magnetic resonance spectroscopy, electron paramagnetic resonance, and impedance spectroscopy [29–31].

During the usage of a Li-ion battery, the major component of Li undergoes reversible de−/intercalation reaction [38, 39]. However, part of Li gets irreversibly lost with the lifecycle of usage. The Li inventory is lost due to the irreversible reactions, consuming Li to form SEI layer (solid electrolyte interface) and typically taking place at the negative electrode competing directly with the reversible intercalation reaction. The growth of SEI during initial cycles is desirable as it helps in protecting the electrode against solvent decomposition. Over the cycling, the irreversible reactions continue forming the SEI layer, and a gradual loss of cyclable Li causes capacity fade.

Solid-electrolyte interphase (SEI) is a thin layer formed at the active material-electrolyte interphase. The SEI is formed by the reduction reaction of the electrolyte forming a thin layer of organic-inorganic salts over the active material. The thickness of the SEI layer ranges between tens to hundreds of nanometers. The exact composition of SEI can be debated, and based on the operating conditions, different researchers have found it to be different. The SEI provides beneficial kinetic stability to the electrolyte against further decomposition, ensuring stable cycling performance. With cycling, the layers of SEI grow on the top of active material leading to incremental capacity fade and passivation to intercalation, although the rate of SEI layer growth slows down after the initial few cycles because of the non-availability of the fresh surface of active material. The mechanics of particle cracking are intrinsically coupled to the SEI layer growth, because the microcrack formation and propagation lead to the availability of fresh active material surface, leading to faster growth of SEI. The inherent coupling between these two major modes of degradation of LIB is explored hereafter.

Methodology

Computational models for Li-plating require accounting for an additional side reaction shown in Eq. (3.54) apart from the model described using Eq. (3.1) to Eq. (3.15):

$$Li^+ + e^- \xrightarrow{\text{Li-plating}} Li \tag{3.54}$$

$$j = j_1 + j_2$$

$$j_1 = k \left(C_s^{0.5} C_e^{0.5} (C_{s,\max} - C_s)^{0.5} \right) \left(\left(\exp \left(\frac{F \eta_1}{2RT} \right) - \exp \left(-\frac{F \eta_1}{2RT} \right) \right) \right)$$

$$\eta_1 = \phi_s - \phi_e - U_1 - j R_{\text{film}}$$

$$j_2 = \min \left(0, k_p C_e^{0.3} \left(\exp \left(\frac{0.3 \eta_2 F}{RT} \right) - \exp \left(-\frac{0.7 \eta_2 F}{RT} \right) \right) \right) \tag{3.55}$$

$$\eta_2 = \phi_s - \phi_e - U_{\text{Li}^+/\text{Li}} - j R_{\text{film}}$$

$$U_{\text{Li}^+/\text{Li}} = 0.0 \ V$$

$$\frac{\partial \delta_{\text{film}}}{\partial t} = -\frac{j_2 M}{\rho F} \qquad R_{\text{film}} = \frac{\delta_{\text{film}}}{k_{\text{film}}}$$

The total reaction current density is composed of two components: Li-plating (j_2) and intercalation (j_1) as shown in Eq. (3.55). Li-plating is assumed to be irreversible, and thus, the reaction current is restricted to the forward direction, and stripping of plated Li is not considered as shown in Eq. (3.54). The intercalation current density at both electrode is based on the Butler-Volmer equation with equal charge transfer coefficients $\left(\alpha = \frac{1}{2} \right)$. Here "$k$" is the kinetic reaction constant for the intercalation; "C_s" and "C_e" represent the solid-phase and electrolyte-phase concentrations. ϕ_s and ϕ_e represent the solid-phase and electrolyte-phase potential; η_1 and η_2 represent the driving overpotentials for the intercalation and Li-plating reaction. The open-circuit potential (U) for graphite and NMC is a material property, function of the surface Li-concentration; it has been adapted from literature [32]. "j_2" shows the Li-plating current density; it is modeled based on the Butler-Volmer-like kinetics with unequal charge transfer coefficients. Equation (3.55) involves a minima function that ensures that Li-plating is triggered only if the overpotential for Li-plating ($\eta_2 < 0$ V) drops below zero as per the work of Ge et al. [29]. Deposition of Li apart from the loss of cyclable lithium can cause the growth of the passivation layer. This leads to the growth of film resistance toward electrochemical reaction, calculated as film thickness divided by film conductivity. The amount of Li-deposition (n_{Li}) is used to quantify the severity of Li-plating in the anode, for this, we sum up the volumetric plating current density within the anode as shown in Eq. 3.56:

$$\frac{dn_{\text{Li}}}{dt} = -\int_0^{L_{\text{anode}}} a_s j_2 A_{\text{cs}} dx \tag{3.56}$$

To model the additional side reaction for SEI formation, we include the irreversible SEI formation reaction kinetics to the above-presented model. That is, now Li$^+$ ions have two reaction pathways, reversible intercalation into the electrode or irreversible SEI layer formation:

$$j_{\text{total}} = j_{\text{SEI}} + j_{\text{I}} \tag{3.57}$$

The reaction current density for intercalation and SEI are given by the Butler-Volmer formulation:

$$j_I = i_{0,I} \left[\exp\left(\frac{F\eta_I}{2RT} \right) - \exp\left(-\frac{F\eta_I}{2RT} \right) \right] \tag{3.58}$$

where the overpotential (η_I) for the reaction is given by Eq. (3.59):

$$\eta_I = \phi_s - \phi_e - U - (R_{SEI} \times j_{total}) \tag{3.59}$$

SEI formation follows the Tafel formulation:

$$j_{SEI} = -i_{0,SEI} \left[\exp\left(-\frac{F\eta_{SEI}}{2RT} \right) \right] \tag{3.60}$$

Here, the overpotential for SEI is defined as:

$$\eta_{SEI} = \phi_s - \phi_e - U_{SEI} - (R_{SEI} \times j_{total}) \tag{3.61}$$

In this equation, ϕ_s and ϕ_e are the solid-phase and electrolyte-phase potentials of the electrode, respectively. U and U_{SEI} are the open circuit potentials for intercalation and SEI formations, respectively. The growth of the SEI layer leads to passivation of the active material interface; the resistance of the layer is given by:

$$R_{SEI} = \frac{\delta(t)}{\kappa_{SEI}} \tag{3.62}$$

Here κ_{SEI} is the ionic conductivity of the SEI layer, and $\delta(t)$ is the thickness of the SEI layer. The rate of growth of SEI volume is given by the following relation:

$$\frac{dV_{SEI}}{dt} = -\frac{j_{SEI}M_{SEI}A_{SEI}}{\rho_{SEI}} \tag{3.63}$$

$$\delta(t) = \frac{V_{SEI}(t)}{A_{SEI}(t)} \tag{3.64}$$

where V_{SEI} is the volume of the SEI formed and thickness is the volume divided by the effective area.

Representative Highlights

The impact of coupled electrochemical-thermal degradation is analyzed for graphite-NMC based Li-ion cells. Figure 3.6a shows a schematic representation of the intercalation and Li-plating reaction occurring at the surface of the anode particles in a typical Li-ion cell system. The efficient operation of the Li-ion

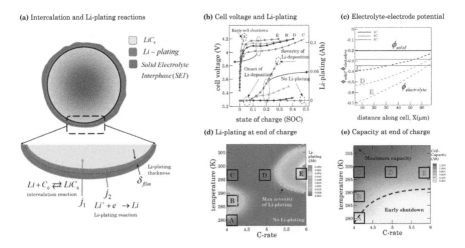

Fig. 3.6 Li-plating degradation during fast charging of LIB's (Reproduced with permission from Ref. [8]. Copyright 2020 American Chemical Society) (**a**) Schematic representation of the electrochemical reactions consisting of intercalation and Li-plating occurring over the graphite particle; (**b–e**) Operation of a LIB under different ambient temperature and C-rate; (**b**) Cell voltage and Li-plating; (**c**) Solid and Electrolyte potentials in anode; (**d**) Li-plating; (**e**) Cell capacity

cell system under a variety of operating conditions is paramount to its diverse application. Since fast-charging and low-temperature conditions are known to accelerate the Li-plating degradation, we herein study the C-rate and temperature ranging between 4C and 6C, 0 °C to 25 °C to gain a mechanistic understanding of Li-plating.

Figure 3.6b marks points A-C for probing the impact of temperature while C-E for investigating the effect of C-rate. The microstructural composition of the anode and cathode has been fixed, with both having a porosity of 25%, leading to a capacity of 2.29 Ah. Fig. 3.6b shows the cell voltage vs state of charge (SOC) for the different operating conditions. Two factors control the severity of Li-plating, the onset of Li-plating and the time of charging of the cell. The Li-plating damage would be maximum when the onset of Li-plating has occurred with the cell undergoing charging for a long time. Other cases with either late onset of Li-plating or shorter charging time avoid the severe Li-plating damage. As we can see that Li-plating from C to E is significantly varying and highly dependent on C-rate (temperature is fixed). From C to E in Fig. 3.6b, we observe that Li-plating increases from zero to 0.075 Ah as the C-rate increases from C to E. This makes the onset of Li-plating occur much earlier; meanwhile, the cutoff state of charge decreases from 0.5 to 0.3. As we go from point C to point E, the increased C-rate of operating increases the reaction rate per unit volume in the anode. The gradients of electrolyte potential and concentration become steeper along the length to provide higher Li-ion flux to the anode as seen in Fig. 3.6c. Even though the capacity of the cell is reduced due to earlier cutoff, the rate of electrodeposition increases. So, we see an early onset of Li-plating at E (6C), although no Li-plating for C (4C) is shown in Fig. 3.6b. The Li-plating severity shows an increasing trend from C to E (0.0 to 0.07 Ah). This also results in a higher heat generation rate from C to E, while they undergo the same temperature rise as seen in Fig. 3.6b due to a higher heat generation rate accompanied by brief charging time (E) or vice versa (C).

Going from C (298 K) to A (278 K), we can observe a decreasing ambient temperature. With the decreasing ambient operating temperature, the physical processes become sluggish in the cell, particularly the temperature-dependent kinetic reaction constant for intercalation reaction. This also results in a higher kinetic component of heat generation in A (278 K) compared to C (298 K). Although this does not directly lead to a rise in cell temperature, because of shorter charging periods from C to A and natural convective cooling. The coupling between temperature and Li-plating is visible for the B, where the rate of Li-plating decreases after 0.15 state of charge. Such a coupling becomes even stronger for the adiabatic fast-charging case because all the heat generated by the cell raises its own temperature and affects the electrochemical processes. The ambient temperature plays a critical role; lowering the temperature increases intercalation resistance by increasing the kinetic overpotential; this also results in the early onset of Li-plating at A, although Li-plating severity shows a nonmonotonic trend, with A (278 K) and C (298 K) experiencing almost no plating and B (288 K) experiencing 0.04 Ah. In a temperature above 298 K (C), the high temperature makes the intercalation reaction rate constant high enough that no onset of Li-plating occurs, while in A (278 K), the capacity of the cell reduces by an order of magnitude due to early cutoff of charge phase.

Li-plating and electrochemical performance of the battery is a strong function of the microstructural configuration of both electrodes. To understand the fast charge capabilities, we keep the cathode microstructure fixed with a porosity of 25% and thickness of 80 μm corresponding to a 2.2 Ah capacity. The anode porosity is increased from 15% to 45% and thickness is increased from 66.42 μm to 94.09 μm to achieve the nominal capacity of 2.2 Ah for the anode and the cathode.

Figure 3.7e shows the useable region for charging under different porosity and C-rate. For quantifying the safety and efficiency of a battery's operation, we define four indicators. Two performance indicators quantify the charging capacity (Q_c) and discharging capacity (Q_d) and two safety indicators quantify the maximum temperature (T) and Li-plating degradation (Q_p). Three regions are defined based on the values of the performance indicator and safety indicator as shown in Fig. 3.7a–d. Charging and discharging indicators should have range (low, 0–0.5 Ah; medium, 0.5–1 Ah; and high, >1 Ah). Cell temperature is designated as safe, < 60 °C; at risk, 60 °C–80 °C; and unsafe, > 80 °C. Plating capacity maps are delineated into safe, < 2.5%; at risk, 2.5%–5%; and unsafe, >5%. An intersection of the attributes based on the four is used for generating the final performance safety map.

From Fig. 3.7a–d, we observe that operating at an ambient temperature of 25 °C, the major limiting factor is the temperature rise and degradation safety. Figure 3.7c shows the temperature can reach the risk-prone or unsafe region if used at higher C-rates between 3 and 5. The Li-plating degradation contour map shows unsafe,

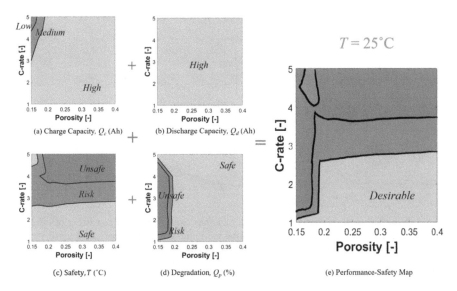

Fig. 3.7 Contour maps [33] for (**a**) charge capacity, (**b**) discharge capacity (low, 0–0.5 Ah; medium, 0.5–1 Ah; and high, >1 Ah), (**c**) cell temperature rise (safe, <60 °C; at risk, 60 °C-80 °C; and unsafe, >80 °C), and (**d**) plating capacity (safe, < 2.5%; at risk, 2.5–5%; and unsafe >5%) as a function of current rate and porosity at ambient temperature 25 °C. (**e**) The intersection of electrochemical performance-safety indicators gives the desirable zone for ambient temperature 25 °C. (Figure reproduced with permission from Ref [33] Copyright (2020) (The Electrochemical Society))

and risk-prone regions are vertically aligned, implying the lower porosity between 15% and 20% porosity anode tends to have significant Li-plating damage. The intersection of these four regions gives us a safe and efficient operating condition. Overall an anode with porosity ranging between 20% and 40% can be operated between C-rate 1 and 2.5. The operation at low porosity and high C-rate is limited by the sluggish kinetics, electrolyte, and solid-phase transport and due to the small temperature rise before cutoff leading to limited temperature rise and marginal property improvements.

Since the SEI growth over the active material and the diffusion-induced particle cracking are coupled to each other, we look at the individual impact of each and then their coupled effect.

Figure 3.8a shows that during delithiation of the particle, microcracks propagate either internally or from the surface. The cracks propagating internally cause diffusivity of the particle to reduce, because of the additional hindrance; on other hand the surface cracks increase the electrochemically available area. From Fig. 3.9a, we see the evolution of the interfacial area of a graphite particle; during the initial half of the discharge process, the total electrochemical surface area is effective, and it increases. During the next half of the discharge process, we see that the electrochemically ineffective area starts to increase, and the effective surface

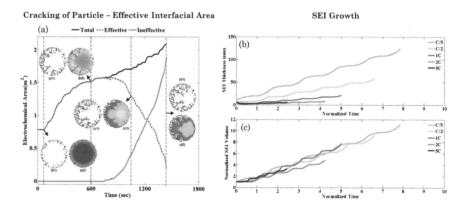

Fig. 3.8 (**a**) The total interfacial area in a particle has two components consisting of effective surface area and ineffective surface area. During the initial discharging period, effective surface area increases because of cracking of particles, while in the second half, depletion of Li makes the surface unable to contribute to Li flux hence the ineffective surface area increases; (**b, c**) Formation of SEI layer occurs when the electrolyte comes in contact with fresh active material surfaces leading to passivation of surface and loss of Li-inventory. The growth of SEI mainly occurs on the anode and during lithiation, i.e., charging of cell causing the wiggles. (**b**) SEI thickness grows highest for lower C-rate because of longer charge and discharges time; (**c**) SEI volume growth occurs almost at the same rate because at a higher rate, small time of charging is compensated by the higher area due to cracking of particle [35]

area starts to decrease. Also, the contribution of ineffective surface area increases significantly in the second half. The depletion of lithium at the surface of the particle in the second half makes the surface area ineffective, and thus effective surface area is only available within the microcracks of the particle.

The evolution of the thickness of the SEI layer with time is shown in Fig. 3.8b for operation at different C-rates. The SEI growth takes place on the anode during the charging process, while no growth occurs during the discharge, giving rise to wiggles. The SEI thickness is obtained by dividing the SEI volume by the effective surface area; thus the growth of the SEI layer takes place on the electrochemically active area. At lower rates, the charge and discharge occur for a long time which provides more time for SEI to grow thicker. For higher rates of charge and discharge, we observe a smaller thickness of the SEI layer, although Fig. 3.8c shows that the growth of SEI volume is almost equal across the different C-rates. There are two competing effects governing the SEI volume, an increase in effective surface area of particle providing a new surface for SEI volume to grow. On the other hand, an increase in area decreases SEI current density (J_{SEI}). These two effects negate each other to a significant extent for the different C-rates studied here.

Figure 3.9 shows the electrochemical performance curves for four models differing in the extent of physics incorporated in each model [35]. The single particle model does not account for fracture effects while only accounting for electrochemistry. The mechanics-only model incorporates additional reduced diffusion transport in particles which leads to worse performance as compared to the previous model.

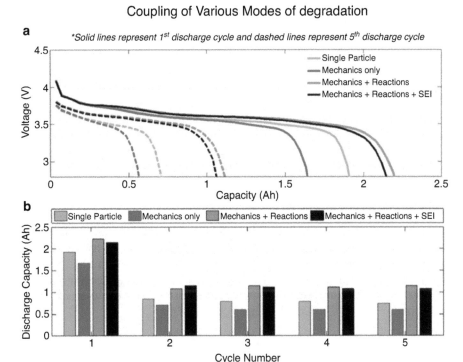

Fig. 3.9 Comparison between four different models: (i) Single-particle model (green bar), (ii) mechanical fracture model (blue bar), (iii) mechanical fracture combined with reaction within the cracks connected to the surface (red bar), and (iv) combined mechanical fracture, the reaction within cracks and SEI growth model (black bar) [35]. (Figure reproduced with permission from Ref [35] Copyright (2018) (The Electrochemical Society))

The mechanics + reactions model additionally accounts for the surface area created by the fracturing of particles, which enhances the reaction kinetics and performs best among all models. Finally, the mechanics + reactions + SEI model additionally accounts for the SEI growth in addition to the above models. Due to passivating effects of SEI over the anode, it leads to additional kinetic overpotential along with loss of cyclable lithium. Looking at the overall degradation modes, we see that stress-induced cracking to a certain extent improves the electrochemical kinetics by providing additional surface area which enhances the performance of the cell.

Intercalation Stress-Induced Mechanical Damage

Charging and discharging of a Li-ion battery involves repeated de−/intercalation out of host material, i.e., graphite/NMC, etc. This along with the diffusion of the Li in the host causes concentration gradients in the particle, resulting in stress generation to achieve mechanical equilibrium. For this reason, they are termed diffusion-induced stresses (DIS) and cause damage initiation and nucleation [34–36]. A parallelism can be drawn between stresses caused due to thermal expansion and diffusion-induced stresses. Thermal gradients $\Delta T(K)$ within the particles lead to the expansion of particles proportional to the coefficient of thermal expansion, $\alpha(K^{-1})$. Similarly, Li concentration gradients, $\Delta c(mol/m^3)$, inside active material particles cause a strain proportional to the molar expansion coefficient Ω (m^3/mol). When particles are constrained from expansion, it results in stress and the storage of stress energy which when it exceeds the fracture threshold energy of the material leads to crack nucleation or propagation. These stresses originating just from the diffusion inside particle are considered, but the presence of secondary phases such as conductive binder additive exacerbates stresses causing additional mechanical damage. The method developed here can predict the mechanical damage due to the diffusion-induced stresses and captures the microcrack evolution and propagation.

Insertion of Li causes volume expansion in the host and leads to strain ranging between a few to the order of a hundred percentage. State-of-the-art graphite anodes exhibit volume expansion of 10%, while high-capacity silicon and tin anodes show about 300% volumetric change. The anode undergoes significant mechanical damage which scales in portions to these volumetric strains and causes cracking and pulverization of active material leading to capacity fade.

Methodology

For simulating the mechanical behavior of the intercalation material, it is represented as a network of springs (see Fig. 3.10). The random lattice spring formalism is coupled with the solid-state diffusion of lithium and electrochemical performance analysis. The computational domain is divided into a spring network with each node connected to six springs. To capture the mechanical properties of the actual material, two distinct types of springs are used, central springs (n) and shear springs (s). The values of spring constants (k_s) and (k_n) are adjusted to replicate Young's modulus (E) and Poisson's ratio (ν) of the material. The relations for Young's modulus and Poisson's ratio in terms of the axial and shear spring stiffness:

$$E(c) = \frac{\sqrt{3}}{4l}\left((3-\nu)k_n + (1+\nu)k_s\right)$$

$$\nu = \frac{k_n - k_s}{3k_n + k_s}$$

(3.65)

To determine the stress and force distribution within the particles, a quasistatic force equilibrium is assumed. Since the Li diffusion inside active particles is a very slow process, assumptions of quasistatic analysis of mechanical equilibrium are sufficient. Body forces are absent, and quasistatic equilibrium implies inertial term

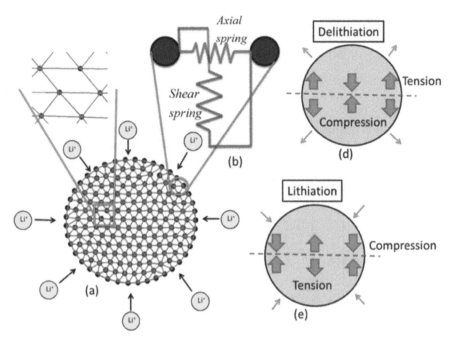

Fig. 3.10 A schematic representation of the lattice spring model was adopted in the current analysis. (**a**) Discretization of the circular domain into springs. Lithium ions intercalate or deintercalates from the outer surface. (**b**) Magnified visualization of each spring. All the mass is assumed to be lumped at each node. The springs show axial as well as shear stiffness. (**c**) A magnified visualization of a broken spring. (**d**) Schematic diagram showing compression close to the central region and tension around the peripheral region during delithiation representatively shown on the cut-plane (dashed line). (**e**) Schematic diagram showing tension close to the center and compression around the periphery during lithiation representatively shown on the cut-plane (dashed line). (Reproduced with permission from Ref. 36. Copyright 2013 IOP Publishing)

is zero. Hence, the stress distribution inside the particle can be obtained by solving for Eq. 3.66.

$$\nabla.\sigma = 0 \qquad (3.66)$$

Boundary condition includes the surface of the active material particle being stress-free. Coupling of the electrochemistry is ensured by using the single-particle model (SPM). Under the assumptions of sufficiently high ionic conductivity of electrolyte, the electrolyte concentration and potential gradients along the electrode thickness are negligible and ignored. The rate-limiting step is the solid-state diffusion inside the active material particles; Li concentration is solved using Fick's diffusion equation:

$$\frac{\partial C_s}{\partial t} = \nabla.(D\nabla C_s) \qquad (3.67)$$

Boundary conditions to the surface of the active particle involve a constant flux from all directions, irrespective of the distance from the separator:

$$-D\frac{\partial C_s}{\partial n} = \frac{i}{F}$$
$$I_{app} = \int\limits_{electrode} i \, dS \tag{3.68}$$

Based on the diffusion equation, the Li concentration inside the active material particle is solved, using which the concentration gradients are calculated. The stress-strain relation is converted to a local force vs displacement relation. The stresses generated due to the Li diffusion are incorporated as axial displacement inside the spring using Eq. (3.69).

$$\begin{Bmatrix} f_{a1} \\ f_{s1} \\ f_{a2} \\ f_{s2} \end{Bmatrix} = \begin{bmatrix} k_n & 0 & -k_n & 0 \\ 0 & k_s & 0 & -k_s \\ -k_n & 0 & k_n & 0 \\ 0 & -k_s & 0 & k_s \end{bmatrix} \begin{Bmatrix} \Omega \Delta c_1 l \\ 0 \\ \Omega \Delta c_2 l \\ 0 \end{Bmatrix} \tag{3.69}$$

The force due to the local displacement is calculated based on Eq. 3.70.

$$\begin{Bmatrix} f_{a1} \\ f_{s1} \\ f_{a2} \\ f_{s2} \end{Bmatrix} = \begin{Bmatrix} f_{a1} \\ f_{s1} \\ f_{a2} \\ f_{s2} \end{Bmatrix}_{mech} + \begin{Bmatrix} f_{a1} \\ f_{s1} \\ f_{a2} \\ f_{s2} \end{Bmatrix}_{conc} = \begin{bmatrix} k_n & 0 & -k_n & 0 \\ 0 & k_s & 0 & -k_s \\ -k_n & 0 & k_n & 0 \\ 0 & -k_s & 0 & k_s \end{bmatrix} \begin{Bmatrix} u_{a1} \\ u_{s1} \\ u_{a2} \\ u_{s2} \end{Bmatrix}$$
$$\tag{3.70}$$

Energy stored in each spring is calculated using $\Psi = \frac{1}{2}\vec{F}_{spring}\vec{u}_{spring}$. If the strain energy of the spring exceeds the breaking energy threshold, it would lead to spring being irreversibly removed from the network, physically meaning fracture of the material.

In the formulation discussed above, the impact of Li diffusion onto particle stress and strain is accounted, but the impact of microcrack on the diffusion process is not accounted. For ensuring the reverse coupling, i.e., the impact of stress on diffusivity is implemented by reducing the diffusivity of the affected region as $D_{eff} = \alpha D_{original}$. The value of the parameter α is taken as 0.6; the value is reasoned based on the work of Barai et al. [36].

The electrochemical performance of the battery is estimated based on the single-particle model formulation based on the work of Guo et al. [37]. This model couples the electrochemical reaction at the particle surface with the diffusion of Li into the particle. The assumption of high enough electrolyte conductivity leads to negligible electrolyte concentration and potential gradients. At high C-rates (> 2) appreciable gradients in the electrolyte form, the present model can be extended by using a C-rate-dependent electrolyte transport resistance (R_e). Butler-Volmer kinetics formulation is used to describe the electrochemical reactions occurring at the anode and cathode electrodes, already described in Eq. 3.4.

3.3.2 Microstructure Evolution and Electrolyte Transport Dynamics in Lithium-Sulfur Batteries

Figure 3.11a presents a representative of carbon-sulfur composite electrode microstructure that consists of 40% carbon and 10% precipitate by volume, which in turn corresponds to a pristine porosity of 60%. Pristine porosity of the cathode refers to the porosity before any precipitation occurs. This is preceding the loading of sulfur content, which would eventually result in a reduction of electrode porosity at the end of cell fabrication. Depending on the interfacial energies of the carbon-

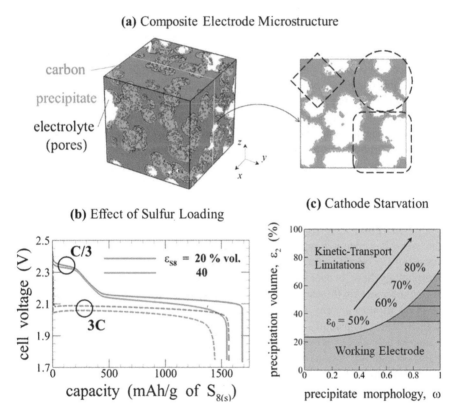

Fig. 3.11 (**a**) Li-sulfur electrode microstructure consisting of 40% carbon and 10% precipitate by volume; (**b**) Effect of sulfur loading on electrochemical performance; (**c**) Identification of cathode starvation zones based on pore blockage and surface passivation effect. (Reproduced with permission from Ref. 45. Copyright 2017 American Chemical Society)

precipitate and precipitate-precipitate interfaces, the Li_2S precipitate can vary from a film-type to a finger-like morphology. The microstructural arrangement of the precipitate governs the manifestation of limiting mechanisms such as pore blockage and surface passivation and alters the buildup of cell resistance during electrochemical operation. Based on a microstructure-coupled electrochemical performance mode [45], Fig. 3.11b presents the effect of sulfur loading on the discharge behavior of Li-S cells for a pristine porosity of 75%, pore radius of 1 μm, and a finger-like precipitate morphology. The electrochemical performance of Li-S battery electrodes depends on the mutually coupled processes of electrode microstructural transformation, transport of various sulfur species in the electrolyte, and the associated interfacial reaction kinetics. As shown in Fig. 3.11b, an increase in sulfur loading results in a faster buildup of cell resistance, leading to a reduced specific cell capacity. For a specific pristine cathode, discharge current increases with a larger sulfur loading and leads to more precipitation interactions for a given interval of time. Consequently, both larger kinetic and transport resistances are exhibited at increased sulfur loading. In addition, electrochemical performance is also affected by the morphology of the precipitate. A more finger-like precipitate morphology would present enhanced active surface area and result in reduced magnitudes of kinetic resistance. On the other hand, film-like morphologies tend to induce lower cell capacities due to the limited availability of active surface area, which results in exacerbated surface passivation limitations. Based on the volume of precipitate and the precipitate morphology, Fig. 3.11c demarcates the regime of kinetic-transport limitations (depicted in pink), which depends on the severity of mechanisms like surface passivation and pore-phase blockage. While film-like morphologies tend to cause surface passivation, finger-like morphologies result in intensive pore-phase blockage. With an increase in pristine porosity, it is observed that the cathode can sustain larger precipitation before it becomes transport-limiting, and the critical precipitation amount required for pore blockage is delayed. Consequently, the joint regime of passivation and blockage-based starvation (pink) is altered by the variation in pristine porosity. In Fig. 3.11c, this porosity-dependent regime is depicted in blue to showcase its effect on cathode starvations. Overall, sulfur-to-electrolyte ratio, cathode porosity, and precipitate morphology have been identified as key microstructural descriptors that have a strong influence on the electrochemical performance of Li-S batteries [45].

In addition to the electrode microstructure, charge carriers in the electrolyte of a Li-S battery constantly evolve and exhibit specification, which alters its ionic transport behavior over cell operation. The electrolyte facilitates charge and species transport and dissolution of polysulfides and hosts interactions between the cathode microstructure and precipitate, playing a pivotal role in the electrochemical performance of the Li-S battery. The evolution of species in the electrolyte is also strongly correlated with the composition, spatial arrangement, and microstructural changes of the cathode. Over cell operation, the impregnated sulfur in a porous carbon structure undergoes successive reduction to form a lithium sulfide precipitate, which is strongly correlated with the speciation and evolution dynamics of the electrolyte. Built on nonequilibrium thermodynamics principles, a concentrated solution theory

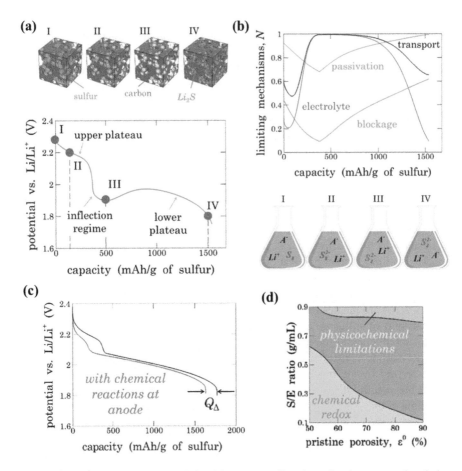

Fig. 3.12 (**a**) Evolution of cell potential and the corresponding electrode microstructural evolution of a carbon-sulfur composite electrode containing 20% vol sulfur and 80% pristine porosity, which is operated at 1C; (**b**) Severity of different limiting mechanisms during electrochemical operation (Reproduced with permission from ref. 47 Copyright 2018 American Chemical Society). (**c**) Effect of chemical redox-based degradation on cell performance; (**d**) Demarcation of viable electrodes based on chemical redox effect and physicochemical limitations. (Reproduced with permission from ref. 48 Copyright 2018 American Chemical Society)

has been developed to capture the progression of electrolyte transport [47], taking into account both inter- and intra-species interactions. For a 1C discharge of a Li-S cathode with 20% sulfur loading (by volume) and 80% pristine porosity, Fig. 3.12a illustrates a representative electrochemical performance signature based on the concentrated solution theory. This depicts two voltage plateaus connected by a regime of inflection. Herein, the first plateau corresponds to the balance between the dissolution of sulfur and reduction to long-chain polysulfides, while the second plateau marks the precipitation of lithium sulfide.

Accounting for the electrolyte transport and electrode microstructure evolution, three limiting mechanisms, namely, electrolyte conduction, pore blockage, and surface passivation, have been captured over cell operation. Both ionic conduction limitation and blockage jointly determine the overall electrolyte transport resistance. As shown in Fig. 3.12b, concurrent tracking of these limiting mechanisms with cell performance reveals that the inflection point in Fig. 3.12a corresponds to a steep increase in electrolyte transport resistance. This is physically associated with a large rise in the concentration of the medium-chain polysulfide during the intermediate stages of discharge that limits the electrolyte's ionic conductivity. Additionally, the physical evolution of the electrode also occurs, exhibiting an initial increase in pore space with the dissolution of the solid sulfur and a subsequent shrinkage due to lithium sulfide precipitation. The effect of these precipitation/dissolution interactions is observed in Fig. 3.12b as blockage and passivation modes of resistance, predominantly during the beginning and end of discharge operation. Speciation in the electrolyte and microstructural evolution are closely related to mechanisms like surface passivation and pore blockage and contribute to the buildup of the cell's internal resistance. Fundamental understanding of the coupled set of electrochemical complexations, species transport, and microstructural interactions is critical toward the design of high-performance Li-S batteries.

Transport of sulfur species in the electrolyte toward the anode and subsequent chemical reduction of these species are the two distinct mechanisms that constitute the "polysulfide shuttle" effect [48]. The chemical reduction at the anode is jointly characterized by the reactivity and concentration of sulfur species. On the other hand, the electrolyte transport behavior is strongly dependent on the physical evolution of the cathode and electrolyte speciation. The two modes of sulfur species transport to the anode are via migration and diffusion, which are correlated to the concentration and electric potential gradients in the electrolyte phase, respectively. Subsequent chemical reductions leading to a loss in capacity can be classified into reversible and irreversible defects. The reduction of intermediate-chain polysulfides that lead to the formation of lithium sulfide results in an irreversible loss since they cannot be oxidized to recover the capacity. On the contrary, the reduction of long-chain polysulfides and dissolved sulfur correspond to a reversible capacity defect. Figure 3.12c presents the joint effect of the reversible and irreversible modes of capacity defect due to polysulfide shuttle on electrochemical performance. Depending on regimes of cell operation, a capacity defect due to the polysulfide shuttle phenomenon can either be reaction-limited or shuttle-limited [48]. Lastly, a classification of viable electrodes based on the pristine porosity and the sulfur-to-electrolyte ratio is presented in Fig. 3.12d, considering the effect of chemical redox and physicochemical limitations. As shown in Fig. 3.12d, while a lower sulfur-to-electrolyte ratio and porosity lead to larger chemical redox limitations, higher sulfur content is associated with limitations pertaining to the chemical and physical evolution of the electrolyte and cathode. An optimal regime of cathode porosity and the sulfur-to-electrolyte ratio is identified in Fig. 3.12d, which involves marginal contributions from both these limitations. A comprehensive understanding of mechanisms underlying the polysulfide shuttle effect and associated cell performance implications would pave the way toward a systematic design of Li-S battery cathodes and electrolytes.

3.3.3 Polymer Electrolyte Fuel Cells

3.3.3.1 Direct Numerical Simulation

In contrast to the volume averaging method adopted in the macrohomogeneous approach, the fine details of the pore structure are well captured in DNS models [56–58]. Thus pore-scale direct numerical simulations (DNS) performed without homogenization over the fully resolved porous structures which are obtained from either experimental imaging techniques or stochastic generation methods is an accurate method to study the influence of the microstructure on the coupled reactive species and charge transport. However, a drawback associated with these calculations is the increase in computational overhead. In addition, the DNS models improve the predictions compared to the traditional macrohomogenous models employing empirical property correlations like the Bruggeman tortuosity factor by incorporating accurate physics-based information.

Experimental imaging and effective property simulations using DNS can be synergistically used to unravel structure-transport-property relationships [59, 60]. The analysis was performed on a reconstructed structure which has been extracted from nanoscale X-ray computed tomography (nano-CT) as seen in Fig. 3.13a. The gray color map indicates carbon, platinum, and primary pores below the nano-CT resolution, while on the other hand, the bright color map conforming to the intensity of Cs^+ ions reveals the presence of ionomer agglomerates. Most of the gray surface has low-intensity Cs^+ surrounding it, thus illustrating the effect of film coverage. In Fig. 3.13b, the corresponding solid and pore size distributions have been portrayed. The solid agglomerates have sizes less than 1 μm and adhere to a normal distribution. Similarly, the pore phase exhibits a log-normal-type distribution with the pore sizes being smaller than 1 μm. The shortest path tortuosity which can be considered as a lower bound for transport tortuosity provides a description of the possible percolation pathways from the entrance to the exit in a complicated porous geometry. In Fig. 3.13c, the distribution of shortest path tortuosity values along with the pathways for both the pore and solid phase have been shown. It can be observed that the pore network shows a higher average tortuous structure when compared to the solid.

3.3.3.2 Degradation Interactions

Large-scale commercialization of PEMFC technology would only be able to see the light of the day when the high cost of the platinum (Pt) catalyst is reduced and durability challenges are circumvented [61, 62]. The durability issues are congruent

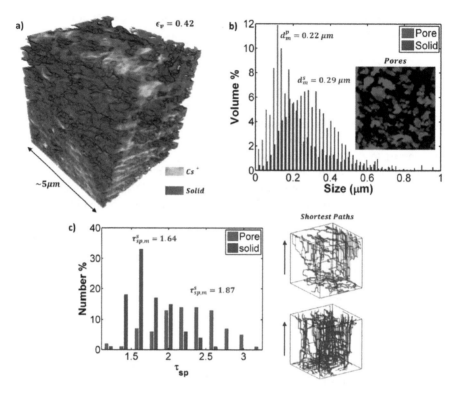

Fig. 3.13 (**a**) A sub-volume upon electrode structural characterization revealing the Cs^+ intensities and the pore morphology, (**b**) corresponding pore and solid size distributions, (**c**) calculated shortest paths and the tortuosity values for both the solid and pore phases. (Reproduced with permission from Ref. 60 Copyright 2017 IOP Publishing)

with the cycle life of the PEMFC when it is exposed to dynamic, start-stop, and idling conditions. Under such a scenario, loss in the electrochemical active area (ECA) and valuable catalyst inventory occurs which impedes the performance of the fuel cell. Concerted efforts have been put to understand the fundamental mechanisms that lead to platinum degradation and to suggest effective mitigation strategies. The relevant degradation phenomena that can appear in an operating automotive environment have been discussed in great detail in recent review articles [63, 64]. A few triggering mechanisms have been identified in this context which has been schematically displayed in Fig. 3.14: (a) Ostwald ripening on a carbon support, (b) Pt coalescence leading to agglomeration, (c) detachment of catalyst particles induced due to carbon corrosion at high reduction potentials, and (iv) Pt dissolution, ion transport through the ionomer phase, and subsequent reprecipitation to form a band in the membrane. Oxidation of the carbonaceous material occurs at high reduction potentials which is deleterious as it may lead to the isolation of Pt particles and microstructural collapse [65, 66] consequently resulting in degradation of the cell performance. In this regard, it is necessary to have models that are cognizant

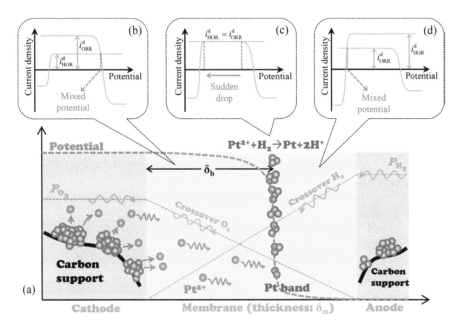

Fig. 3.14 (**a**) Schematic of the degradation mechanisms at play that leads to the formation of Pt band in the membrane followed by the diffusion-controlled mixed potential analysis of the HOR and ORR currents, (**b**) toward the cathode side of the Pt band, (**c**) at the location of the Pt band, and (**d**) toward the anode side of the Pt band. Here, p_{O_2} and p_{H_2} are the partial pressures of oxygen and hydrogen, respectively; δ_b represents the location of the Pt band from the membrane-CCL interface, and δ_m is the membrane thickness; i^d_{HOR} and i^d_{ORR} are the diffusion-limited current densities of the HOR and ORR, respectively. (Reproduced with permission from ref. 64 Copyright 2020 Elsevier)

of the degradation phenomena that can aid in making informed decisions. Such computational capabilities can then be scaled with additional models simulating electrochemistry, water, and heat management to obtain rich information about multiple transport processes that occur in tandem within a catalyst layer.

Methodology

The simulation domain that is utilized while modeling catalyst degradation is pseudo-two-dimensional in nature. It comprises the transport direction in x (denoted by subscript i) having N control volumes and discrete diameter bins in y (denoted by subscript j) where the control volumes are denoted by M. This means that there exist M diameter groups in every i location (x-direction) along with the catalyst layer. The diameter in each particle group is designated by $d_{i,j}$, while the fractional oxide coverage is denoted by $\theta_{i,j}$.

As a consequence of long-term cycling, there is a change in the size of the particle diameters and the reversible formation/removal of the oxide coverage. The transient evolution of the same is governed by Eqs. 3.71 and 3.72:

$$\frac{d\left(d_{ij}\right)}{dt} = -r_{\text{net},\text{Pt}^{2+}} \Omega_{\text{Pt}} \tag{3.71}$$

$$\frac{d\left(\theta_{ij}\right)}{dt} = \frac{r_{\text{net,oxide}}}{\Gamma} - 2\frac{\theta_{ij}}{d_{ij}}\frac{d\left(d_{ij}\right)}{dt} \tag{3.72}$$

The relevant thermo-kinetic forward and backward reaction rates are an extension of Butler-Volmer kinetics by accommodating the effect of the Gibbs-Thomson approximation to cater to the instability of particle diameter.

In addition, the diffusion of dissolved Pt^{2+} ions through the ionomer phase is given by:

$$\varepsilon_{\text{ionomer}}\frac{\partial c_{\text{Pt}^{2+}}}{\partial t} = \nabla \cdot \left(\varepsilon_{\text{ionomer}}^{1.5} D \nabla c_{\text{Pt}^{2+}}\right) + S_{\text{Pt}^{2+}} \tag{3.73}$$

$$S_{\text{Pt}^{2+}} = \sum_{j=1}^{M} \frac{\frac{\pi}{2}\left(d_{i,j}\right)^2 \text{Num}_{i,j} r_{\text{net},\text{Pt}^{2+}}}{N} \tag{3.74}$$

where D is the diffusivity of Pt^{2+} in the ionomer. A coupled solution of Eqs. (3.73 and 3.74) is sought to obtain the change in the number density, particle diameter, and, finally, the electrochemically active area (ECA) within the catalyst layer.

Representative Highlights

In order to achieve a robust performance of the catalyst layer, interactions between the complex material phases need to be ameliorated [67]. The thin catalyst layer is intrinsically heterogeneous, and as such, it is quite difficult to probe the multiple phenomena occurring inside it. Over the recent years, there has been substantial advancement in the imaging techniques like focused ion beam scanning electron microscopy (FIB-SEM) and X-ray tomography which have aided in revealing the convoluted arrangement of the microstructure. In this context, one constraint that existed was in the form of the difficulty in identifying the ionomer layer over the catalyst particles. The remedy for this drawback was obtained through a judicious blend of a computational mesoscale formalism and coupling it with data obtained from tomography experiments [68]. The ionomer coverage over the platinum-carbon interface was found to be heterogeneous in nature, and there existed an opportunity to modulate it to achieve suitable electrochemical performance. Figure 3.15a shows the pore and solid phases for representative volume elements (RVEs) which were reconstructed from FIB-SEM tomography data at varying stages of carbon corrosion (top row). The ionomer network is then resolved based on a physics-based description once the arrangement of the pore and solid phases is

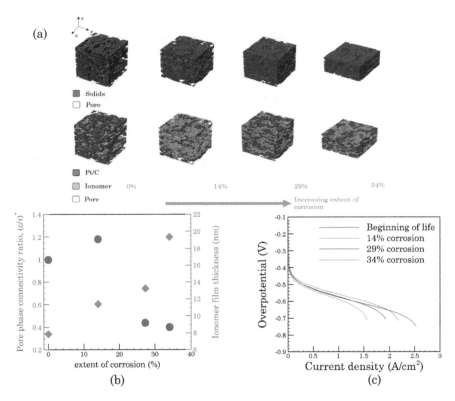

Fig. 3.15 (**a**) Representative volume elements (RVEs) comprising of the pore and solid skeleton have been reconstructed based on experimental FIB-SEM tomographic information (top row). The composite electrodes with the desired amount of ionomer phase (green color) (bottom row), (**b**) the evolution of nominal pore-phase connectivity and ionomer film thickness with the progression of carbon corrosion, (**c**) polarization curves for electrodes at different extents of corrosion [68]. (Figure reproduced with permission from Ref [68] Copyright (2020) (The Electrochemical Society))

obtained. The samples where the ionomer has been added are shown in the bottom row of Fig. 3.15a.

Once the structural arrangement of the composite microstructure was known, it aided in understanding carbon corrosion which leads to oxidation of carbonaceous material thereby inducing the collapse of the electrode material. Accordingly, several reconstructed structures at increasing extents of corrosion were considered, and their structure-transport-property relationships were extracted.

To study the influence of carbon corrosion on the resulting electrochemical response, the effective microstructural properties pertaining to pristine and corroded electrodes were computed as seen in Fig. 3.15b [68]. It was observed that closed or isolated pores develop in the microstructure as and when it is exposed to an aging landscape. A descriptor referred to as pore phase connectivity is defined which is the ratio of the porosity to tortuosity of the electrode. It was further normalized with the reference being the connectivity of the beginning-of-life electrode (BOL). Figure 3.15b indicates the increased severity in the convolution of the pore network since the nominal pore connectivity decreases as the corrosion event progresses. Further, due to the increment in the ionomer to carbon weight ratio; a thicker ionomer film is formed. This in turn leads to elevated effective ionomer conductivity in the aged catalyst layers. Once the structural-transport-property relationships were abstracted, the performance could then be eventually probed as seen in Fig. 3.15c. Observing the ohmic regime highlights that an immediately corroded electrode (14%) behaves better in comparison to a fresh electrode with no corrosion due to better transport of protons owing to its enhanced corresponding conductivity. However, oxygen transport is set to be rate-limiting because of the formation of dead pores along with the prevalence of a bulkier ionomer layer which leads to early onset of mass transport effects with increasing extent of corrosion.

3.3.3.3 Two-Phase Models

Lattice Boltzmann method (LBM) [50, 69, 70] is a mesoscopic technique based on a bottom-up approach of solving the conservation equations by upscaling from a formulation based on kinetic equations. It has been utilized for the purpose of simulating multiphase-multicomponent flows in porous architectures and fluid-structure interactions and also to capture interfacial phenomena. Instead of a continuum, this method assumes the fluid as being a collection of pseudo particles which are then dictated by streaming and collision steps. Due to its inherent parallel nature, LBM is amenable to high-performance computing and has significantly garnered the attention of researchers in the field of computational fluid dynamics.

Representative Highlights

LBM can be employed to understand the effect of the by-product of the ORR, that is, liquid water on the reaction area passivation and pore site blockage in the CCL and GDL microstructures. In Fig. 3.16a, a two-phase LBM model was used to study the flooding effect as a consequence of liquid water formation in a reconstructed CL microstructure [69]. From this analysis, the two-phase statistics in the form of relative permeability can be obtained as a function of the local saturation. These relationships can then be harnessed and fed into macroscopic models to delineate the signature of the electrochemical response of the CCL in the mass-transport regime.

The 3D distribution of the liquid water at increasing extents of the level of saturation (less than 15%) within the CCL under steady-state conditions is depicted in Fig. 3.16a. Below the 10% saturation limit, the percolation pathways for the liquid water are not evident in the CCL microstructure, and accordingly, there is negligible relative mobility of the liquid water with respect to the air phase. With the increase in saturation, a well-connected percolation path is established owing to the effect of capillary forces. In Fig. 3.16b, the 2D liquid water saturation maps at several

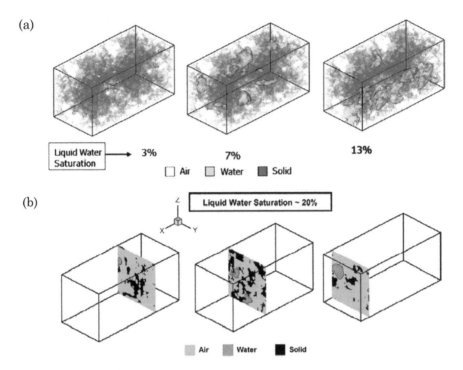

Fig. 3.16 (**a**) 3D liquid water distribution in a reconstructed CCL microstructure at varying saturation levels, (**b**) 2D slices of the phase distribution at different locations in a reconstructed CCL microstructure for a representative saturation of 20%. (Reproduced with permission from ref. 69 Copyright 2009 Elsevier)

locations along the thickness direction of the CCL with 20% saturation are shown. The cross-sectional maps indicate the reduction in the active interfacial area due to the filling of the geometry with liquid water. This in turn highlights the deleterious effects of flooding in the form of kinetic loss which can be evaluated via accurate pore-scale models.

From the context of two-phase dynamics, the hysteresis between imbibition and drainage mechanisms can be modeled by simulating mercury intrusion porosimetry (MIP) at different capillary pressures in the transport direction of the CCL geometry [70]. For such a simulation scenario, constant pressure is set at the inlet boundary; no flux boundary condition is set at the outlet with periodic boundary conditions in the other orthogonal directions. The calculations are performed till the system reaches a steady state. Once it is achieved, the previous saturation profile is employed for the upcoming intrusion stage and then allowed to reach steady-state conditions which are continued till 10 MPa. The drainage process of the curve is obtained in a reverse manner wherein the liquid leaves from the geometry back into the film. The difference between the two curves is referred to as capillary hysteresis. There is a demarcation between the two curves at lower capillary pressures as seen in

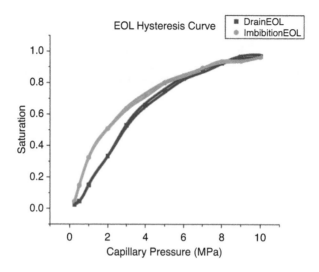

Fig. 3.17 Hysteresis between drainage and imbibition curves for a range of capillary pressures from 0.25 MPa to 10 MPa. The catalyst layer geometries were reconstructed from FIB-SEM tomography data. (Reproduced with permission from Ref. 70 Copyright 2021 IOP Publishing)

Fig. 3.17. As a general feature, it can be observed that there exists a rapid rise in saturation (during imbibition) at lower capillary pressure which is followed by an asymptotic behavior under fully saturated conditions.

3.4 Summary and Outlook

A detailed understanding of the mesoscale complexations that occur within the diverse electrochemical systems illustrated in the previous sections is of prime importance to achieve improvement in its performance and durability. Mesoscale modeling plays an overarching role as illustrated in Fig. 3.18 by interconnecting spatiotemporal scales [71] and is hence crucial in advancing the science and engineering of electrochemical energy devices.

The multiscale and multiphysics nature of the physicochemical interactions has been investigated through several computational and experimental techniques. The porous reactors are very thin; the in situ probing of the participating mechanisms is generally cumbersome. In addition, this also circumvents the availability of input data for the various mesoscale models. However, of late, the advent of high-resolution and non-destructive visualization techniques like focused ion beam scanning electron microscopy (FIB-SEM) and neutron imaging, etc. has made resolution of the convoluted structure of the electrode microstructure plausible. These recent improvements have laid the platform for detailed mechanistic analysis through a synergy of experimental and mesoscale modeling methodologies. A

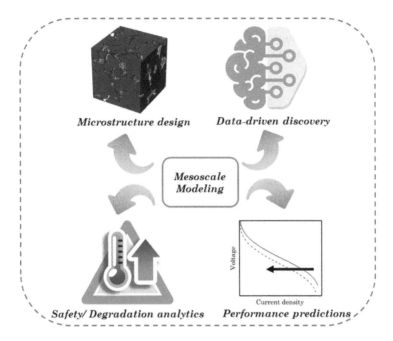

Microstructure design   *Data-driven discovery*

Mesoscale Modeling

Safety/ Degradation analytics   *Performance predictions*

Fig. 3.18 Overarching role of mesoscale modeling paradigm

representative example is advancing the understanding of the flooding dynamics in the resolved geometries of a PEM Fuel Cell cathode by mapping the transients of water distribution obtained through X-ray tomography experiments with accurate mesoscale models.

In fact, predictions made by mesoscale models can also facilitate material discovery and design. High-throughput screening and molecular dynamics simulations have been extensively used to cater to topical areas such as stable and safe electrolyte discovery for energy storage systems like lithium-ion, sodium-ion, and magnesium-ion batteries. A notable effort in this computational framework includes the Materials Genome Initiative [72] and pymatgen library [73] that can significantly aid in boosting the pace of materials discovery. Electrocatalytic studies can be complemented with mesoscale models as well in order to assess the functionality and performance metrics of novel catalysts. This is of prime relevance in the adoption of PEFCs in the electrification domain by expanding its market range from passenger to heavy-duty vehicles [61]. In addition to this, the mesoscale modeling paradigm could serve as an asset in the rational design of electrode architectures that can yield a superior electrochemical system response. This includes the design of energy-dense electrodes in lithium-ion batteries by incorporating through-channels for facile ion travel in the transport direction [74]. Further, recent research areas related to the creation of fuel cell cathodes with accessible carbon support to negate ionomer poisoning can include a judicious mix of experimental-computational techniques [75, 76].

Safety is another critical domain wherein results from mesoscale modeling can be fruitful in the advancement of lithium-ion batteries. The thermal response of lithium-ion batteries under abuse conditions is manifested in the form of internal short circuit, thermal runaway, and separator melting and may finally result in a catastrophic failure. A multiscale model that leverages information from a microstructure scale to a cell-format level such as a pouch cell or a cylindrical cell could help in understanding thermal signatures and help delineate strategies to suppress the triggering mechanisms. A comprehensive CFD-based thermal propagation model can also be utilized to ensure the critical aspect of safety of the battery packs.

The traditional macrohomogenous models were employed by ignoring the pore-scale features of the electrode microstructure. Although this limitation has been overcome through the advanced DNS models, the computational cost associated with running the simulations poses a challenge. The electrochemical sciences community can gain impetus from the mature CFD field by making use of robust solvers with preconditioning capabilities and the advantage of parallelizability. In this direction, machine learning can play an exceedingly crucial role in accelerating the pathway of scientific discovery and decreasing the computational demand at the same time. This can be achieved by building an infrastructure of reduced-order models through an intelligent blend of physics-based and data-driven (statistical) models [77–80].

Acknowledgments Financial support in part from National Science Foundation (NSF grant: 1805215) is gratefully acknowledged. The authors acknowledge the American Society of Mechanical Engineers, American Chemical Society, Elsevier, the Electrochemical Society, and the Royal Society of Chemistry for the figures reproduced in this chapter from the referenced publications of their journals.

References

1. Liu Y, Zhu Y, Cui Y (2019) Challenges and opportunities towards fast-charging battery materials. Nat Energy 4(7):540–550
2. Choi JW, Aurbach D (2016) Promise and reality of post-lithium-ion batteries with high energy densities. Nat Rev Mater 1(4):1–16
3. Wagner FT, Lakshmanan B, Mathias MF (2010) Electrochemistry and the future of the automobile. J Phys Chem Lett 1(14):2204–2219
4. Whittingham MS (2004) Lithium batteries and cathode materials. Chem Rev 104(10):4271–4302
5. Armand M, Tarascon J-M (2008) Building better batteries. *Nature* 451(7179):652–657
6. Ahmed S et al (2017) Enabling fast charging–a battery technology gap assessment. J Power Sources 367:250–262
7. Fear C et al (2021) Mechanistic underpinnings of thermal gradient induced inhomogeneity in lithium plating. Energy Storage Mater 35:500–511
8. Kabra V, et al (2020) Mechanistic analysis of microstructural attributes to lithium plating in fast charging. ACS Applied Materials & Interfaces 12(50):55795–55808

9. Taiwo OO et al (2017) Investigation of cycling-induced microstructural degradation in silicon-based electrodes in lithium-ion batteries using X-ray nanotomography. Electrochim Acta 253:85–92
10. Zielke L et al (2015) Three-phase multiscale modeling of a $LiCoO_2$ cathode: combining the advantages of FIB–SEM imaging and x-ray tomography. Adv Energy Mater 5(5):1401612
11. Nelson GJ et al (2017) Transport-geometry interactions in Li-ion cathode materials imaged using X-ray nanotomography. J Electrochem Soc 164(7):A1412
12. Ebner M et al (2013) X-ray tomography of porous, transition metal oxide based lithium ion battery electrodes. Adv Energy Mater 3(7):845–850
13. Mistry A et al (2016) Analysis of long-range interaction in lithium-ion battery electrodes. J Electrochem Energy Convers Storage 13(3):37
14. Ji Y, Zhang Y, Wang C-Y (2013) Li-ion cell operation at low temperatures. J Electrochem Soc 160(4):A636
15. Mistry AN, Smith K, Mukherjee PP (2018) Secondary-phase stochastics in lithium-ion battery electrodes. ACS Appl Mater Interfaces 10(7):6317–6326
16. Mistry AN, Smith K, Mukherjee PP (2018) Electrochemistry coupled mesoscale complexations in electrodes lead to thermo-electrochemical extremes. ACS Appl Mater Interfaces 10(34):28644–28655
17. Chen C-F, Verma A, Mukherjee PP (2017) Probing the role of electrode microstructure in the lithium-ion battery thermal behavior. J Electrochem Soc 164(11):E3146
18. Vetter J et al (2005) Ageing mechanisms in lithium-ion batteries. J Power Sources 147(1–2):269–281
19. Barré A et al (2013) A review on lithium-ion battery ageing mechanisms and estimations for automotive applications. J Power Sources 241:680–689
20. Broussely M et al (2005) Main aging mechanisms in li ion batteries. J Power Sources 146(1–2):90–96
21. Yoshida T et al (2006) Degradation mechanism and life prediction of lithium-ion batteries. J Electrochem Soc 153(3):A576
22. Chen C-F, Barai P, Mukherjee PP (2016) An overview of degradation phenomena modeling in lithium-ion battery electrodes. Curr Opin Chem Eng 13:82–90
23. Petzl M, Kasper M, Danzer MA (2015) Lithium plating in a commercial lithium-ion battery–a low-temperature aging study. J Power Sources 275:799–807
24. Fan J, Tan S (2006) Studies on charging lithium-ion cells at low temperatures. J Electrochem Soc 153(6):A1081
25. Senyshyn A et al (2015) Low-temperature performance of li-ion batteries: the behavior of lithiated graphite. J Power Sources 282:235–240
26. Smart MC, Ratnakumar BV (2011) Effects of electrolyte composition on lithium plating in lithium-ion cells. J Electrochem Soc 158(4):A379
27. Bugga RV, Smart MC (2010) Lithium plating behavior in lithium-ion cells. ECS Trans 25(36):241
28. von Lüders C et al (2017) Lithium plating in lithium-ion batteries investigated by voltage relaxation and in situ neutron diffraction. J Power Sources 342:17–23
29. Ge H et al (2017) Investigating lithium plating in lithium-ion batteries at low temperatures using electrochemical model with NMR assisted parameterization. J Electrochem Soc 164(6):A1050
30. Wandt J et al (2018) Quantitative and time-resolved detection of lithium plating on graphite anodes in lithium ion batteries. Mater Today 21(3):231–240
31. Schindler S et al (2016) Voltage relaxation and impedance spectroscopy as in-operando methods for the detection of lithium plating on graphitic anodes in commercial lithium-ion cells. J Power Sources 304:170–180
32. Valøen LO, Reimers JN (2005) Transport properties of LiPF6-based Li-ion battery electrolytes. J Electrochem Soc 152(5):A882
33. Vishnugopi BS, Verma A, Mukherjee PP (2020) Fast charging of lithium-ion batteries via electrode engineering. J Electrochem Soc 167(9):090508

34. Verma P, Maire P, Novák P (2010) A review of the features and analyses of the solid electrolyte interphase in li-ion batteries. Electrochim Acta 55(22):6332–6341
35. Kotak N et al (2018) Electrochemistry-mechanics coupling in intercalation electrodes. J Electrochem Soc 165(5):A1064
36. Barai P, Mukherjee PP (2013) Stochastic analysis of diffusion induced damage in lithium-ion battery electrodes. J Electrochem Soc 160(6):A955
37. Guo M, Sikha G, White RE (2010) Single-particle model for a lithium-ion cell: thermal behavior. J Electrochem Soc 158(2):A122
38. Christensen J, Newman J (2003) Effect of anode film resistance on the charge/discharge capacity of a lithium-ion battery. J Electrochem Soc 150(11):A1416
39. Christensen J, Newman J (2004) A mathematical model for the lithium-ion negative electrode solid electrolyte interphase. J Electrochem Soc 151(11):A1977
40. Fang X, Peng H (2015) A revolution in electrodes: recent progress in rechargeable lithium–sulfur batteries. Small 11(13):1488–1511
41. Hagen M, Hanselmann D, Ahlbrecht K, Maça R, Gerber D, Tübke J (2015) Lithium–sulfur cells: the gap between the state-of-the-art and the requirements for high energy battery cells. Adv Energy Mater 5(16):1401986
42. Wild M, O'neill L, Zhang T, Purkayastha R, Minton G, Marinescu M, Offer G (2015) Lithium sulfur batteries, a mechanistic review. Energy Environ Sci 8(12):3477–3494
43. Zhang SS (2013) Liquid electrolyte lithium/sulfur battery: fundamental chemistry, problems, and solutions. J Power Sources 231:153–162
44. Bruce PG, Freunberger SA, Hardwick LJ, Tarascon J-M (2012) Li–O 2 and li–S batteries with high energy storage. Nat Mater 11(1):19–29
45. Mistry A, Mukherjee PP (2017) Precipitation–microstructure interactions in the li-sulfur battery electrode. J Phys Chem C 121(47):26256–26264
46. Chen C-F, Mistry A, Mukherjee PP (2017) Probing impedance and microstructure evolution in lithium–sulfur battery electrodes. J Phys Chem C 121(39):21206–21216
47. Mistry AN, Mukherjee PP (2018) Electrolyte transport evolution dynamics in lithium–sulfur batteries. J Phys Chem C 122(32):18329–18335
48. Mistry AN, Mukherjee PP (2018) "Shuttle" in polysulfide shuttle: friend or foe? J Phys Chem C 122(42):23845–23851
49. Cano ZP et al (2018) Batteries and fuel cells for emerging electric vehicle markets. Nat Energy 3(4):279–289
50. Mukherjee PP, Kang Q, Wang C-Y (2011) Pore-scale modeling of two-phase transport in polymer electrolyte fuel cells—progress and perspective. Energy Environ Sci 4(2):346–369
51. Meng H, Wang C-Y (2005) Model of two-phase flow and flooding dynamics in polymer electrolyte fuel cells. J Electrochem Soc 152(9):A1733
52. Pasaogullari U, Wang C-Y (2005) Two-phase modeling and flooding prediction of polymer electrolyte fuel cells. J Electrochem Soc 152(2):A380
53. Pasaogullari U, Wang C-Y (2004) Two-phase transport and the role of micro-porous layer in polymer electrolyte fuel cells. Electrochim Acta 49(25):4359–4369
54. Wang Y, Wang C-Y (2006) A non-isothermal, two-phase model for polymer electrolyte fuel cells. J Electrochem Soc 153(6):A1193
55. Weber AZ, Darling RM, Newman J (2004) Modeling two-phase behavior in PEFCs. J Electrochem Soc 151(10):A1715
56. Mukherjee PP, Wang C-Y (2006) Stochastic microstructure reconstruction and direct numerical simulation of the PEFC catalyst layer. J Electrochem Soc 153(5):A840
57. Wang G, Mukherjee PP, Wang C-Y (2006) Direct numerical simulation (DNS) modeling of PEFC electrodes: part I. regular microstructure. Electrochim Acta 51(15):3139–3150
58. Mukherjee PP, Wang C-Y (2007) Direct numerical simulation modeling of bilayer cathode catalyst layers in polymer electrolyte fuel cells. J Electrochem Soc 154(11):B1121
59. Cetinbas FC et al (2017) Hybrid approach combining multiple characterization techniques and simulations for microstructural analysis of proton exchange membrane fuel cell electrodes. J Power Sources 344:62–73

60. Cetinbas FC et al (2017) Microstructural analysis and transport resistances of low-platinum-loaded PEFC electrodes. J Electrochem Soc 164(14):F1596
61. Cullen DA et al (2021) New roads and challenges for fuel cells in heavy-duty transportation. *Nature*, Energy:1–13
62. Weber AZ, Kusoglu A (2014) Unexplained transport resistances for low-loaded fuel-cell catalyst layers. J Mater Chem A 2(41):17207–17211
63. Prokop M, Drakselova M, Bouzek K (2020) Review of the experimental study and prediction of Pt-based catalyst degradation during PEM fuel cell operation. Curr Opin Electrochem 20:20–27
64. Ren P et al (2020) Degradation mechanisms of proton exchange membrane fuel cell under typical automotive operating conditions. Prog Energy Combust Sci 80:100859
65. Star AG, Fuller TF (2017) FIB-SEM tomography connects microstructure to corrosion-induced performance loss in PEMFC cathodes. J Electrochem Soc 164(9):F901
66. Young AP, Stumper J, Gyenge E (2009) Characterizing the structural degradation in a PEMFC cathode catalyst layer: carbon corrosion. J Electrochem Soc 156(8):B913
67. Grunewald JB et al (2019) Perspective—mesoscale physics in the catalyst layer of proton exchange membrane fuel cells. J Electrochem Soc 166(7):F3089
68. Goswami N et al (2020) Corrosion-induced microstructural variability affects transport-kinetics interaction in PEM fuel cell catalyst layers. J Electrochem Soc 167(8):084519
69. Mukherjee PP, Wang C-Y, Kang Q (2009) Mesoscopic modeling of two-phase behavior and flooding phenomena in polymer electrolyte fuel cells. Electrochim Acta 54(27):6861–6875
70. Grunewald JB et al (2021) Two-phase dynamics and hysteresis in the PEM fuel cell catalyst layer with the lattice-Boltzmann method. J Electrochem Soc 168(2):024521
71. Ryan EM, Mukherjee PP (2019) Mesoscale modeling in electrochemical devices—a critical perspective. Prog Energy Combust Sci 71:118–142
72. National Science and Technology Council (US). Materials genome initiative for global competitiveness. Executive Office of the President, National Science and Technology Council, 2011
73. Ong SP et al (2013) Python materials genomics (pymatgen): a robust, open-source python library for materials analysis. Comput Mater Sci 68:314–319
74. Amin R et al (2018) Electrochemical characterization of high energy density graphite electrodes made by freeze-casting. ACS Appl Energy Mater 1(9):4976–4981
75. Yarlagadda V et al (2018) Boosting fuel cell performance with accessible carbon mesopores. ACS Energy Lett 3(3):618–621
76. Cetinbas FC et al (2019) Effects of porous carbon morphology, agglomerate structure and relative humidity on local oxygen transport resistance. J Electrochem Soc 167(1):013508
77. Mistry A, Mukherjee PP (2019) Deconstructing electrode pore network to learn transport distortion. Phys Fluids 31(12):122005
78. Finegan DP, Cooper SJ (2019) Battery safety: data-driven prediction of failure. Joule 3(11):2599–2601
79. Gayon-Lombardo A et al (2020) Pores for thought: generative adversarial networks for stochastic reconstruction of 3D multi-phase electrode microstructures with periodic boundaries. npj Comput Mater 6(1):1–11
80. Kench S, Cooper SJ (2021) Generating three-dimensional structures from a two-dimensional slice with generative adversarial network-based dimensionality expansion. Nat Mach Intell:1–7

Chapter 4
Development of Computer Aided Design Tools for Automotive Batteries

Taeyoung Han and Shailendra Kaushik

Abstract To accelerate the development of safe, reliable, high-performance, and long-lasting lithium-ion battery packs, the automotive industry requires computer-aided engineering (CAE) software tools that accurately represent cell and pack multi-physics phenomena occurring across a wide range of scales. In response to this urgent demand, General Motors assembled a CAEBAT Project Team composed of GM researchers and engineers, ANSYS Inc. software developers, and Professor Ralph E. White of the University of South Carolina and his ESim staff. With the guidance of NREL researchers, the team collaborated to develop a flexible modeling framework that supports multi-physics models and provides simulation process automation for robust engineering. Team accomplishments included clear definition of end-user requirements, physical validation of the models, cell aging and degradation models, and a new framework for multi-physics battery cell, module, and pack simulations. Many new capabilities and enhancements have been incorporated into ANSYS commercial software releases under the CAEBAT program.

4.1 Background

This project had two main tasks, namely, cell- and pack-level model developments. The principal objective of each task was to produce an efficient and flexible simulation tool for prediction of multi-physics battery response under various vehicle operating conditions. In partnership with DOE/NREL, the Project Team has interacted with the CAEBAT working groups to identify end-user needs and established requirements, integrated and enhanced existing battery sub-models, developed cell-level and pack-level software tools, and performed experimental

T. Han · S. Kaushik (✉)
General Motors, LLC, Detroit, MI, USA
e-mail: shailendra.kaushik@gm.com

This is a U.S. government work and not under copyright protection in the U.S.; foreign copyright protection may apply 2023
S. Santhanagopalan (ed.), *Computer Aided Engineering of Batteries*,
Modern Aspects of Electrochemistry 62,
https://doi.org/10.1007/978-3-031-17607-4_4

119

testing to validate the tools. The emphasis in the software integration was to provide a flexible array of modeling choices that can support wide range of applications while enabling a controllable balance between model fidelity and computational cost. The integrated battery design tools capture the relevant physics including electrochemical, thermal, fluid, and structural response, focusing on the intra-cell and inter-cell nonuniformities that critically impact battery performance and life. Under the cell-level task, ESim developed a life model based on an equivalent circuit model (ECM) including a SEI formation branch in the circuit. The semiempirical parameters were obtained from the cell cycle-life testing performed by GM under various operating conditions. GM engineers generated the test database for validation of the nominal heat source model, as well as cell-level and pack-level electrical and thermal performance. ANSYS provided user-defined interfaces that utilize sub-models developed by NREL that can represent multiple active materials and multiple particle sizes and shapes in the electrodes. ANSYS also created interfaces to enable these new tools to interface with current and future battery models developed by others.

At the pack level, the integrated battery design tools have been significantly advanced by the development of innovative reduced-order models, derived and calibrated from the cell-level models and carefully validated through experimental data. ANSYS incorporated the latest advances in battery modeling research, such as the MSMD modeling approach that was developed by NREL. With a focus on ease of use for nonexperts, workflows for system simulation and robust design optimization were automated. GM generated the test database for the physical validation of a 24-cell module and a production-intent battery pack including electrical and thermal performance. GM validated the tools, obtaining satisfactory agreement with the test data. GM also successfully demonstrated thermal runaway simulations for a battery pack having an internal short circuit. With the expected rapid deployment of these design tools to the industry, the project results will accomplish the key goal of accelerating the pace of battery innovation and development for future electric-drive vehicles.

4.2 Introduction

Cell-level and pack-level design tool developments were two main focus areas for the GM CAEBAT project. The project team in consultation with NREL identified and established end-user requirements, augmented pre-existing sub-models, automated existing workflows, and, finally, performed experimental testing to validate the tools. The team also worked on the task of creating interfaces that would enable these tools to seamlessly interact with all current and future battery models.

In the case of cell-level design tool development, the focus was on accurately predicting the multi-physics response of a large-format lithium-ion battery cells. Both automakers and battery suppliers are fully aware of the importance of thermal management of the battery packs. Temperature plays a significant role in its performance, safety, and service life, and hence, the automakers are understandably paying more attention to innovative thermal management ideas to reduce temperature fluctuations and maintain its temperature around an ideal level around 25° C. Both air and liquid heating/cooling thermal management systems are currently being used by automakers. However, depending on which system is used, the severity of the nonuniform temperature distribution can vary. Regardless, nonuniform temperature in a pack can lead to electrically unbalanced modules, which in turn can lead to loss of performance and reduction in effective cell life. It becomes critical for the automakers to ensure minimum levels of temperature variations across each cell and between cells in a pack. If the temperature is too low, there is a risk of reduced power/energy extractable and cell damaging phenomena called lithium plating. If the temperature is too high, it can lead to shortened cell life. The project aimed to develop a cell-level model that can provide a seamless connection among electrochemical, thermal, and structural responses. Equipped with the proposed cell-level tool, the engineers can now assess the impact of their design parameters on the battery's overall performance, i.e., electric power, capacity, safety, state-of-charge, state-of-life, and internal imbalance within a cell due to spatial variations in temperature and current density. Although one-dimensional electrode-scale models are useful in understanding the physics such as electrochemical kinetics, lithium diffusion and transport, and charge conservation and transport in a cell, a three-dimensional model is required to better understand the spatial variations of temperature and current densities in a pack to design advanced cooling/heating thermal management strategies. The project team has collaborated with NREL for the scale coupling between 1D electrode-scale models and the 3D cell model.

Battery service life represents one of the greatest uncertainties in the total lifecycle cost of electric-drive vehicles. The complex dependency of the cell capacity on time, temperature, voltage, number of charge/discharge cycles, electrode microstructure, and depth of discharge is not fully understood and is often neglected in cell models. To simulate cell aging and degradation, the electrode-scale predictions can be enhanced by introducing models that account for capacity fade due to mechanical degradation caused by thermal and mechanical stress and loss of active material due to film formation. Under the cell level task, ESim developed a practical life model based on an ECM including a SEI formation branch in the circuit model. The semiempirical parameters were validated from the cell cycle-life test data performed by GM engineers under various operating conditions.

On the other hand, in the case of the battery pack model development, the focus is primarily on improving battery pack's thermal management system. Its main goal is to maintain the average temperature of cells in a pack within an optimal temperature range between 25° C and 35° C under various environments ranging from −40° C to 50° C. The battery thermal management system (BTMS) should therefore be capable of achieving these objectives with an active BTMS. However, an active BTMS can be challenging to achieve these objectives in terms of cost, power, weight, and volume. With proposed battery pack tools, one can evaluate various cooling/heating strategies easily and efficiently for trade-off design studies. This would also mitigate the need for very expensive hardware build-and-test iterations and replace it with pack-level simulations.

To generate common cell and pack geometries, one can effectively use existing CAD models. For this purpose, ANSYS's Workbench with capability to modify CAD and mesh is used. It can handle complex 3D geometries of battery cells, including current-collector tabs, encapsulation materials, and neighboring cooling and structural details.

To solve for electrochemical-thermal-fluid behavior of a cell or a module of cells, it is required to fully discretize the model comprising of cells and coolant channels and perform full field simulations in order to provide spatial variation of temperature and current densities in all the cells in a module. However, it is currently almost impractical computationally to scale it to pack level with the computational resources available to the automakers. In the future, with advancements in computational hardware, especially with GPUs, and AI/ML-based reduced-order-modeling techniques, the task of providing optimal BTMS that usually requires many pack-level simulations may become feasible.

With computationally inexpensive alternatives to accomplish rapid acceleration of battery-pack innovation, they also rely on lumped mass representation of the cells in a pack for both electrical and thermal simulations using equivalent-circuit models (ECMs). Lumped mass representation of large-format cells in a pack may not have the necessary spatial resolution sufficient for understanding temperature and current density variations within a cell and effectively guide pack design. In order to combine the best features of both 1D (lumped-mass) and 3D approach of full-field simulations, the project proposes co-simulation using ANSYS Simplorer for 1D and ANSYS Fluent CFD package for 3D. Additionally, with the help of innovative reduced order modeling (ROM) techniques developed during the project, the computationally expensive full-field simulations were replaced in the system-level simulations with the realization of fast-running system-level simulations without sacrificing accuracy.

GM leveraged the math model validation process developed by National Institute of Statistical Sciences (NISS) to rigorously verify and validate cell-level tools, thereby assessing their accuracy and usefulness. Physical verification of the nominal heat-source model and cell-level and pack-level electrical and thermal performance was accomplished using a test database that GM generated during the project. Data from thermal characterization with known heat source and boundary conditions was then used in simulating the battery pack under many TMS (thermal management system) operating conditions. The enthalpy difference between the incoming and the outgoing fluid streams in the pack is used to estimate the amount of total heat generation during the pack usage. Individual cell thermal properties and other electrical and electronic components specific to the pack are used to estimate pack-level specific heat and thermal conductivity. To validate the pack's simulation performance, a number of temperature sensors were placed in every module of the pack. Overall, both cell-level and pack-level validations were successfully performed during this project using the measured data.

Successful development of both the cell-level and pack-level design tools are essential in the design of next-generation battery cells and packs for electric vehicles. In this project, the state-of-the-art in pack-level design tool made significant progress via implementation of reduced order models (ROM). These ROMs were derived and calibrated using cell-level models and careful validation with the experimental data. All the latest research and developments in cell-level and pack-level tools have been incorporated into ANSYS's Workbench that offers unsurpassed ease of use and workflow automation for robust design optimization to its users. To meet the main objective of this project, which is to accelerate the pace of battery development for future electric vehicles, it is essential to have accurate and fast-running battery simulation tools, both at the cell and pack levels.

4.3 Battery Simulation Technology Development

The primary target is automotive battery development or CAE engineers who are experts in neither battery physics nor simulation technologies. This community places a high value on process automation and ease of use. A secondary goal is to provide specialists with a convenient drill-through access to expert features such as electrochemistry sub-model details, numerical solution controls, and ROM algorithms. The intent has been to provide flexible support for a range of roles, from electrode- component researchers, to cell designers, to battery pack manufacturers, to EDV system integrators.

The ANSYS Battery Design Tool (ABDT) was developed to automate and integrate the ANSYS tools to make the various components emulate a single vertical application for batteries, with plug-and-play cell and pack capabilities. With appropriate simplifications, the most important design question being addressed is to *predict and optimize the service life of an automotive lithium-ion propulsion battery.*

In this context, cell temperature is the most important prediction, and the top-priority scenarios or use cases for ABDT are listed here in order of increasing complexity:

1. Cell single-point analysis (rapid cell-qualification study).
   The goal of this workflow is to perform single-point (sometimes referred to as lumped or zero-dimensional) cell electrochemistry calculations without heat transfer, as a rough first estimate to support cell specification setting and preliminary design optimization.
2. Cell cooling strategy analysis (cell-level CFD analysis)
   This scenario represents the main cell-level capability based on field simulation with computational fluid dynamics (CFD).
3. Pack network flow analysis
   This is a single-physics scenario to determine the quasi-steady flow distribution among several cells within a pack (or a module), based on fluid-manifold geometry and coolant flow rate. This run is a prerequisite for the transient pack-

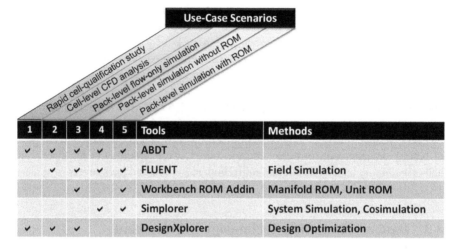

Fig. 4.1 Simulation tools and methods

level simulations, if the flow nonuniformity is significant and impractical to (a) prescribe in advance, e.g., due to time variation, or (b) calculate via CFD cosimulation of the manifold as part of the pack-level simulation.

4. Pack-level simulation without ROM

This workflow is completely Simplorer-based, using system simulation models without any recourse to field simulation other than possibly prior "tuning" of particular model coefficients. This basic scenario serves as a prerequisite for the instantiation of ROMs.

5. Pack level simulation with ROM

This is the most complex scenario and the one that demonstrates the most significant breakthrough for new CAEBAT-developed technology. It builds on the workflows outlined in the other blocks above.

Figure 4.1 summarizes the mapping between the use cases, methods, and software building blocks.

4.3.1 Field Simulation Based on MSMD Approach

The challenge of thermal-electrochemical coupled battery modeling comes from its multi-scale, multi-physics nature. In a lithium-ion (Li-ion) battery, anode and cathode are usually made of active materials coated on the surface of metal foils. The two foils are usually separated by a polymer separator. A battery cell is manufactured by winding or stacking up the sandwich layer into a jelly roll or prismatic shape. A physics-based battery model may require discretization to resolve the anode-separator-cathode sandwich layers or even the position within individual electrode particles. Depending on the goals and constraints of the analysis, it may be prohibitively expensive to resolve all those layers explicitly even for a single battery cell, let alone for battery packs used in electric vehicles.

To handle the complex interplay of multi-physics in this multi-scale environment, Kim et al. developed the multi-domain multi-scale (MSMD) method [1]. In the MSMD method, different physics are resolved at different scales and in different domains. Then they are intimately coupled by exchanging information from one domain to another. One key idea in the MSMD method is that a battery cell can be considered as continuous anisotropic porous media and its electric behavior can be represented by two co-located potential fields. Effective transport properties of the porous media are obtained by averaging the properties of electrolyte and active particles. The dual potential fields are then solved at the same mesh as the temperature, and the source terms of the two potential equations and the energy equation are computed through the subscale model. In such a simulation, the mesh only needs to resolve the thermal scale (rather than each sandwich layer), making the MSMD approach feasible for both battery cell and pack simulations. We implemented the MSMD framework into the ANSYS Fluent solver in a manner that applies to both cylindrical and rectangular/prismatic cell form factors. Three electrochemical sub-models were implemented, and the method was extended to battery pack simulations as well.

4.3.1.1   MSMD Model Framework

Although the positive and the negative electrodes occupy multiple, separate physical regions in reality, in the MSMD method, they are assumed to occupy the entire battery's active zone at the same time. Effective material properties are used so that the average thermal and electric field are correctly predicted. Two potential equations, one for the potential field in the positive current collector and the other for the potential field in the negative current collector, along with the temperature equation, are solved at the battery's dimension scale. The energy balance equation and the electric potential equation are as follows:

$$\frac{\partial \rho C_p T}{\partial t} - \nabla \cdot (k \nabla T) = \dot{q}$$
$$\nabla \cdot (\sigma_+ \nabla \phi_+) = -j_{ECh} \qquad (4.1)$$
$$\nabla \cdot (\sigma_- \nabla \phi_-) = j_{ECh}$$

where T is the temperature; φ_+ and φ_- are the electric potentials for the positive and negative current collectors, respectively; ρ and C_p are the effective density and the specific heat capacity, respectively; and σ_+ and σ_- are the effective electrical conductivities for the positive and negative current collectors. Note that all these material properties are effective values for the active-zone sandwich, which are evaluated by averaging known properties of the constituent materials. Finally, j_{ECh} is the transfer current, and $\dot{q}$ is the internal heat generation rate during operation.

$$\dot{q} = \sigma_+ |\nabla \phi_+|^2 + \sigma_- |\nabla \phi_-|^2 + \dot{q}_{ECh} \tag{4.2}$$

The first two terms in Eq. (4.2) are the contributions of the Ohmic heating from the positive and negative current collectors. The third term is the contribution from the electrochemical reactions.

The boundary conditions for the potential equations are as follows:

$$\begin{aligned} \phi_- = 0, \ n \cdot \nabla \phi_+ = 0 \text{ at negative tab} \\ n \cdot \nabla \phi_- = 0 \text{ at positive tab} \end{aligned} \tag{4.3}$$

where n is the normal unit vector. The boundary condition for φ_+ at the positive tab is defined depending on various operating conditions

$$\begin{aligned} \varphi_+ &= V_{tab} & \text{if voltage is specified} \\ A_p \, n \cdot (-\sigma_+ \nabla \varphi_+) &= I_{tab} & \text{if current is specified} \\ \varphi_+ \left[A_p \, n \cdot (-\sigma_+ \nabla \varphi_+) \right] &= P_{tab} & \text{if power is specified} \\ \varphi_+ / \left[A_p \, n \cdot (-\sigma_+ \nabla \varphi_+) \right] &= R_{tab} & \text{if resistance is specified} \end{aligned} \tag{4.4}$$

where A_p is the battery's positive tab area and V_{tab}, I_{tab}, P_{tab}, and R_{tab} are battery's tab voltage, current, power, and external electric resistance, respectively, at which the battery operates. The external resistance boundary condition can be conveniently used to study an external short circuit.

Although the thermal energy equation is solved over the entire computational domain, including cooling channels in which the fluid flows, the two electric potential equations need to be solved only over the battery's active zones where electrochemical reactions occur and the passive zones where the current can flow but no electrochemical reactions take place – such as battery tabs and bus bars, as shown in Fig. 4.2. It is noteworthy that transfer current density j_{ECh} is equal to zero in these passive zones. Also, note that at the cell level, in the active zones, the transfer current density j_{ECh} and electrochemical heat q_{ECh} are unknown and must be provided from an electrochemical sub-model.

Beside many advantages of MSMD, modularized structure is an important one. This means that one can plug-and-play with any electrochemistry sub-model depending upon the application need. Figure 4.2 shows the coupling between the large-scale model and the subscale electrochemistry model. Each mesh cell in a CFD model is assumed to act as a mini battery in this MSMD method. At each time step during the CFD simulation, local temperature and electric potential fields are passed on to the electrochemistry sub-model to compute the source terms j_{ECh} and q_{ECh}.

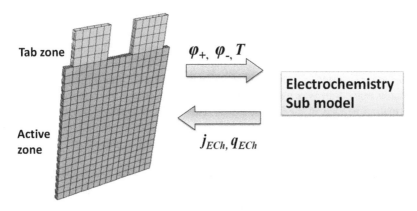

Fig. 4.2 Scale coupling in the MSMD method

A wide range of electrochemical models, with a varying balance between fundamental physics and empiricism, is available in the open literature. The following three electrochemical sub-models are implemented in ANSYS Fluent [2]:

- NTGK model [3–5].
- ECM model [6].
- Newman P2D model [1, 7–9].

In addition, a user-defined E-Chem model option is available for expert users. Expert users can hook up their own implementation of any electrochemical sub-model into ANSYS Fluent's battery model framework.

4.3.1.2   Coupling Between CFD Model and Subscale Electrochemistry Models

During computation, an electrochemical sub-model is constructed and solved in every CFD cell during every flow iteration (for steady simulations) or time step (for transient simulations). For the ECM model, this results in solving a set of equations consisting of one algebraic and three differential equations. For the Newman model, this can involve solving a system of hundreds of differential and algebraic equations, depending upon the number of discretization points used in the electrode and particle domain. This adds to a tremendous computational cost to the simulation.

In fact, the current density is rather uniform. As a result, instead of solving the sub-model for every computational cell, we can group CFD cells into clusters and solve the sub-model only once for each cluster.

4.3.1.3 Battery Pack Model

In many industrial applications, battery cells are connected in an $nPmS$ pattern where n batteries are connected in parallel as a stage and m such stages are connected in a series to form either modules to form a pack or directly form a pack. A search algorithm was used to automatically determine the battery connectivity information needed in the Fluent solver. This enables the user to provide only the information regarding the battery's active zones, tabs, bus bars, and system tabs, and the solver automatically determines how each cell is electrically connected to its neighboring cells.

Essentially, two electric potential equations are solved in an MSMD framework. If the simulation involves only one battery cell, the positive electric potential φ_+ can be seen as potential field in the positive current collector. Similarly, negative electric potential φ_- can be interpreted as the potential field in the negative current collector. However, this becomes complicated in a battery pack. Consider an example, as shown in Fig. 4.3, in which three cells are connected in series. In this case, a cell's positive tab is connected to the neighboring cell's negative tab resulting in approximately similar levels of positive electric potential of the first cell and negative electric potential of the second cell, which means that they both must be solved on the same domain. Since this is not possible, these electric potentials cannot be used in the battery pack simulations. Also, note that only one potential is solved for passive zones such as tabs and bus bars.

Interestingly, the two potential fields, φ_1 and φ_2, can still be used to capture the electric field for a battery pack that they cannot be interpreted as potential fields at the current collectors anymore. Instead, the same potential variable could be the potential in the positive current collector for one battery and the potential in the negative current collector for the next battery, depending on the battery's connectivity.

Different colors (red and blue) are used in Fig. 4.3 to show different equations being solved. By introducing the concept of "level" of a zone, it becomes convenient

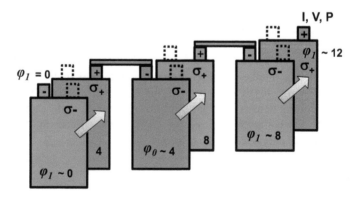

Fig. 4.3 Solution of potential fields in battery packs

for bookkeeping purposes to mark which equation will be solved by which zone. In the case of an active zone, there are two "levels," and hence, both φ_1 and φ_2 are solved. Whereas in the case of passive zones, such as tabs, bus bars, etc., they belong to only one level, and hence only one potential equation is solved. φ_1 is solved for the zones at odd level, and φ_2 is solved in the zones at even level. If, for example, m stages of batteries are connected in series, there will be $m + 1$ "levels." Each level is isolated in the sense of no flux transfer across its boundaries, except for the first and the last levels in a pack. Potential fields of the adjacent levels are coupled only through the equation's source terms. For the convenience of coding, two electric potential equations are solved continuously across all the conductive zones. However, zero electric conductivity is used for the passive zone to mimic that only one potential equation is solved there.

4.3.1.4 Battery Connections in a Pack

Cells in a pack can be connected through either "real connections" or "virtual connections." Real connections are bus bars that are physically resolved and meshed in the model. Virtual connections are bus bars or battery tab volumes that are not explicitly resolved in the model. Instead, the connection information is provided in a battery connection definition file. The solver reads this information and sets the electric boundary conditions for each individual cell through the virtual connections.

Using virtual connections saves the time and effort of meshing bus bar and tab volumes, which are usually very thin and difficult to mesh. However, electric resistance and Joule heating are not considered in these unresolved volumes.

4.3.1.5 Cell Clustering

Although the MSMD method is very effective, there are some challenges preventing the method from being used for large problems in a battery pack simulation. One challenge is to couple a complex electrochemical sub-model such as Newman's P2D model in a three-dimensional thermal simulation. Several partial differential equations already need to be solved in the electrochemical sub-model. If the sub-model is called by every CFD computational cell, the high computational cost makes the method practically unfeasible even in a single cell simulation, let alone battery packs [10]. Guo and White noticed that, under a battery's normal operation, electrochemical reactions progress evenly within a battery and they proposed a linear approximation model [11]. By using that assumption, the computational cost can be reduced by orders of magnitude, thus making the complex sub-model more feasible in a real three-dimensional electrochemical-thermal coupled simulation.

ANSYS Fluent provides a tool for grouping the CFD cells into clusters. For a user-specified number of clusters N_x, N_y, and N_z in the x, y, and z directions, respectively, the solver divides the cell's electrochemical domain into $N_x N_y N_z$ clusters and then obtains the solution for each cluster using the cluster's average values. If $Nx = N_y = N_z = 1$, the Fluent model is basically reduced to the linear approximation model.

A test problem was run and published [10], and it shows that to run one time step in a single cell simulation, it only takes about a second to call the P2D model if the cell clustering approach is used, while it may take hours if the fully coupled method is used.

4.3.1.6   Electric Load Profile

In a battery simulation, the electric load type could be defined by current, voltage, power, or external electrical resistance. All these electric load types have been implemented in the cell model. The user can fix an electric load type and electric load value in a simulation, or he can change either or both load types and values in a simulation on the fly.

4.3.1.7   Parameter Estimation for the NTGK and ECM Models

The NTGK and ECM models are highly empirical in nature. To use them in a simulation, the user needs to provide model parameters, i.e., Y and U functions for the NTGK model parameters, and R_1, R_2, C_1, C_2, R_s, and V_{ocv} are the parameters for the ECM model. These functions are usually battery specific. Fluent provides a parameter estimation tool for computing these parameters from standard battery testing data. For the NTGK model, the testing data are voltage response curves at different discharging C-rates. For the ECM model, the experimental data are hybrid pulse power characterization (HPPC) testing data.

4.3.2   System Simulation

The full CFD simulation of the battery pack is impractical in many cases. This is so since the size of the required finite-volume mesh for battery packs is usually quite large. Field simulation of a battery module can potentially take weeks even using clusters with high-performance computing (HPC) capabilities. In addition, most battery packs are comprised of repeating hierarchical structures such as sections, modules, cells, and/or cooling channels. In some pack designs, foams are placed between some of the cells within the module. This situation makes system modeling based on the domain decomposition or "divide-and-conquer" technique a very useful approach.

4.3.2.1 System Simulation Strategy

The CAEBAT system modeling approach builds upon the concept of divide-and-conquer combined with segregation of physics as illustrated in Fig. 4.4.

For system simulation, ANSYS developed a layered software approach to balance automation and flexibility. This approach is analogous to the cell-level approach with mesh templates and ABDT. The user has a highly automated, intuitive interface for building and solving a system-level model of a battery pack using ANSYS system simulation tool Simplorer, with the option to represent selected items in the pack using CFD models and/or ROM derived from CFD.

The basic building block for the pack, the unit model (see Fig. 4.4), is retrieved from the User Library. It should be emphasized that any configuration of cells, cooling channels, and other items can be created by users and stored to the User Library as a new template. In order to be automated by the Simplorer-ABDT scripts, these models must conform to the conventions explained below in Sect. 4.3.2.2.

In general, the Simplorer-ABDT scripts require that the unit has four ports as sketched in Fig. 4.5. Two ports are the most positive and negative electrical terminals for the overall unit, and two are for the thermal paths between adjacent units. The pack-level simulation does not solve for fluid pressure drop, and therefore the unit model does not involve any hydraulic ports. The scripts will automate the distribution of the coolant flow based on the lookup table and user-prescribed total pack flow rate.

Table 4.1 summarizes the common mechanisms causing cell-to-cell temperature variation in a pack and how the simulation tools in ABDT can address them.

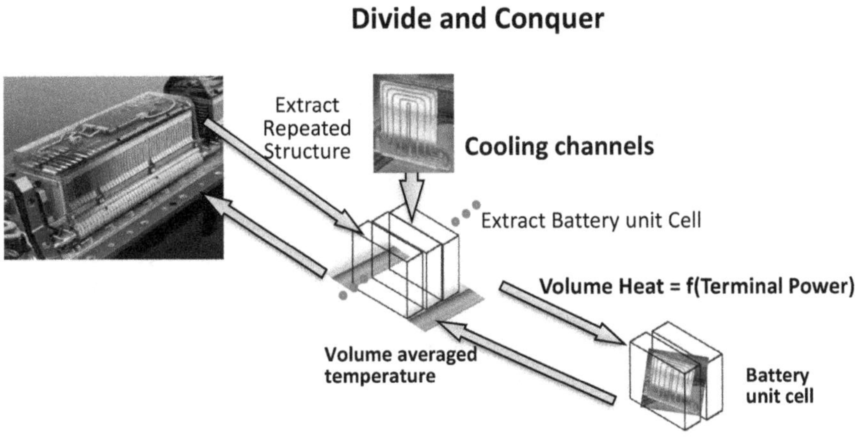

Divide and Conquer

Extract
Repeated
Structure

Cooling channels

Extract Battery unit Cell

Volume Heat = f(Terminal Power)

**Volume averaged
temperature**

**Battery
unit cell**

Fig. 4.4 System modeling strategy

Fig. 4.5 Representation of
the unit model component
block

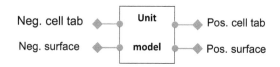

Table 4.1 Mechanisms causing pack temperature variations

Cause of variation in cell-average T	How modeled in CAEBAT pack-level tool
Nonuniform coolant flow distribution – Either inherent in manifold design or due to foreign-object blockage of micro-channel(s)	Included in pack model via flow-distribution table. Can be highly automated via scenario 3
Non-symmetric heat transfer at the ends of each module or other structure within the pack	A separate "end unit" model can be created and instantiated after automated pack creation, replacing the regular symmetric units
Statistical variations or discrete defects from cell manufacturing (ranging from subtle to severe) leading to off-nominal electrical resistance and/or heat generation rate	Possible to include in pack model via manual updates to selected cell data after automated pack creation
Imperfect cell balancing between discharge cycles, with off-nominal SOC creating different heat generation rates	Cell-specific initial conditions including SOC can be input via manual updates after automated pack creation

4.3.2.2   Definition and Configuration of the Unit Model

System simulation rests fundamentally on the principle of domain decomposition, and in this project the "unit" refers to the basic element within the battery pack that can be repeatedly instanced in such a "divide and conquer" strategy. The unit also enables CFD, ROM, and system simulation to be blended together for a flexible balance between rigor and speed. Therefore, the definition of the unit is of vital importance.

Although there are no limitations in principle, the development of the initial CAEBAT User Library focused on battery configurations of immediate interest to GM, to provide the most useful tools and demonstration. The first decision has therefore been to study linear arrays of prismatic cells, where it is possible to define distinct "channels" for fluid cooling "between" the cells – that is, along their dominant "side" areas. Even within this category, a challenge arises in the two kinds of connectivity among cells in such a battery. The physical layout, which determines the fluid/thermal paths, is generally independent from the electrical cell connectivity. Both may be significant in multi-physics simulation.

To address this challenge and demonstrate the flexibility of the methods, the "stub library" being provided as a starting point for further user customization contains the two-unit templates illustrated schematically in Fig. 4.6.

For convenience in applying the manifold-ROM strategy for flow distribution, each unit models a single fin/fluid channel, denoted (f) in the cross-sectional sketches. The one-cell unit is applicable when a simple c-f pair repeats (typical in air-cooled packs). The two-cell unit is common in liquid-cooled packs such as

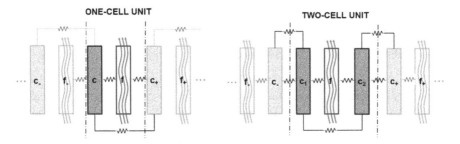

Fig. 4.6 Unit templates contained in the "stub library"

the first-generation Chevy Volt, where two cells (c) share each fin and the pattern c-f-c, c-f-c exists in the array. Portions of the adjacent units are indicated with "-" and "+" labels and grayed out in the figures. The resistors below are thermal, not electrical, with the extra cell-to-cell wires indicating bus bar conduction paths.

The GM test module used for validation (see Sect. 4.4.2) uses 12 pairs of parallel-wired cells (2P12S), and so the 2P wiring scheme was adopted for the example two-cell unit.

With the above considerations and choices in mind, the following assumptions are adopted for the Simplorer pack-creation script:

1. A four-tier hierarchy (unit/group/module/pack) is supported. If the pack is not divided into separate modules (or if a single module is being studied), the number of modules can be set to one.
2. Each unit may contain any number of cells but, as noted above, a unit always contains one channel.
3. Each group is comprised of identical units electrically in parallel.
4. Each module is comprised of identical groups electrically in series, and modules themselves are also in series.
5. By using the manifold ROM, each unit (and therefore each channel) may have a unique total coolant mass flow rate automatically applied. However, for simplicity, the channel inlet fluid temperature is a global time-dependent parameter, assumed the same for each cell.
6. Coolant flow is completely parallel throughout the pack – that is, each particle of fluid travels through only one channel during its passage through the pack.

With these assumptions, the connectivity input from the user simplifies to three integers:

- n_M, number of modules in the pack,
- n_S, number of groups in series within a module,
- n_P, number of units in parallel within a group.

The shorthand notation for the pack in terms of the user input is then.

$n_P P n_M n_S S$ (using the one-cell unit)

or $2 n_P P n_M n_S S$ (using the two-cell unit)

These conventions preserve flexibility in pack configuration via the script while insulating the pre-configured units in the library to changes in both electrical and thermal connectivity. The assumptions above may be considered temporary limitations to be removed in a future version or through user customization. An important note is that, after script execution, even greater flexibility exists using manual Simplorer modeling. As a prime example, the combination of the c-f-c and c-f stored unit models facilitate the creation of configurations having one more cell than fin (or vice versa) by manually adding one special "end-unit" to an automatically generated array.

To help automate the pack-level workflow, the scripts control the creation of some custom GUI panels (via standard Windows. NET libraries, not shown) to accept battery-specific input. The ABDT scripts are written in the IronPython language. The IronPython interpreter and the standard C++ Simplorer user interface (UI) take turns accessing and assembling the model ("design") that defines the pack-level equations to be solved by the Simplorer solver.

4.3.2.3   Building the Unit Model Templates

Several ways exist to implement the unit model templates, and each has its advantages. Using the standard Basic Elements library (e.g., resistor, capacitor, etc.) in Simplorer makes maximum use of the robust, hard-wired C code for computational efficiency, and in principle it reduces code development. An alternative is to hand-code each component using VHDL-AMS language. Although nominally execution may be slower, it offers the possibility to shed the variables and calculations in the standard library components that are not needed for CAEBAT applications. Given the analytical formulas of the ECM and anticipated extensions thereof, this approach also facilitates the explicit coding of time derivatives that may be needed to handle strong nonlinear dependencies of standard quantities such as resistance on state of charge. The Basic Elements VHDLAMS library in Simplorer offers a good compromise between these two options, in that the standard library is available as a starting point, but the source code is exposed, allowing customization as needed.

The Simplorer Thermal library is sufficient to add the thermal behavior to the ECM. In other words, instead of a fictitious second "circuit" that mimics the thermal sub-model, the script uses VHDLAMS Library/Physical Domains Components/Thermal Components. This relieves the script from translating input quantities according to the thermal-electrical analogy, since the solver can natively recognize temperature and heat flux quantities and the associated properties.

Electrical Sub-Model

The electrical sub-model developed for the unit model templates is based on the same ECM implemented in Fluent for cell-level simulation and is taken from the concept proposed by Chen and Rincon-Mora [6]. This sub-model blends the attributes of Thevenin, impedance, and runtime models for accurate prediction of both V-I curves and transient response. It is summarized in [6] and assumes that the cell's self-discharge timescale is much longer than that of the simulation. The negative and positive terminals of the cell are arranged left-to-right. Two RC pairs are fitted to characterize short- and long-term transient behavior. The cell capacitor in the embedded circuit does the "coulomb counting" (i.e., it integrates the cell current via the defined ammeter) to track SOC, which is then used as an input to expressions within all of the other circuit elements, including the open-circuit voltage source.

In general, the values of all circuit elements sketched above are functions of temperature, discharge rate, SOC, and state of health (SOH). In [6], which focused on low-rate, isothermal applications for low-power electronics, all dependencies except for SOC were neglected. The model matches the Fluent implementation of the 21 published SOC curve-fit coefficients documented in [6] for a 0.85 Ah cylindrical Li-ion cell at 25 °C and discharge rates below 1 °C. These fits are highly nonlinear; for example, open-circuit voltage $U(SOC)$ involves an exponential term plus a third-order polynomial. GM experimental results have provided updated values and expanded functional relationships that are more appropriate for large-format cells and vehicle applications. In the meantime, the unit model approximates a larger 24.3 Ah cell by simply scaling up the capacitance by the ratio of the cell capacity compared to that in the publication.

Augmenting this electrical sub-model is a battery heat-source equation presented in [12]. Adopting a cell-integrated basis as is appropriate for system simulation, the rate or total heat generation rate within the cell, Q, is the sum of an irreversible voltage-difference term and a reversible (entropic) term

$$Q = I \left(U - V - T_c \frac{\partial U}{\partial T_c} \right) \tag{4.5}$$

where

I = current (positive for discharge, negative for charge)
U = open-circuit voltage (based on cell-average SOC)
V = cell voltage
T_c = spatial average cell temperature

and other terms are as defined in [6]. The entropic heating term associated with lithium intercalation is not well understood theoretically but is known to be a function of SOC for common cell chemistries. Historically it has often been neglected in practice, but it can be significant at low charge/discharge rates when the

	SOC	dU/dT_c (mV/K)
Table 4.2 Lookup table for entropic heating term	1.00	0.22
	0.82	0.22
	0.56	0.05
	0.47	−0.31
	0.35	0.14
	0.28	0.14
	0.23	−0.03
	0.0	−0.30

magnitude of $(U\text{-}V)$ is relatively small. To provide a starting point and a convenient template for users to include material-specific thermodynamics in the future, the following lookup table (see Table 4.2) is included in ABDT based on experimental measurements of an undoped $Li_yMn_2O_4$ spinel cathode [12].

In the two-cell template (see Fig. 4.6), the electrical sub-model is simply duplicated for each of the two cells, which are then connected in parallel to the unit's two electrical ports.

Thermal Sub-Model

The thermal sub-model of the single-cell unit adopts a simple two-lump approach that is common as an example in heat transfer texts (e.g., [13]). A standard heat-exchanger method [13] is used to relate the mass-averaged temperature of the composite fin/fluid region, T_f, with the coolant temperatures at the channel inlet and outlet:

$$T_o - T_i = \epsilon \left(T_f - T_i\right) \tag{4.6}$$

where effectiveness

$$\epsilon \equiv 1 - \exp\left(-\frac{h_f A_f}{\dot{m} c_o}\right) \tag{4.7}$$

In the above formula, heat-exchanger effectiveness is the ratio of actual heat transfer to the maximum possible heat transfer. h_f is the average convection heat transfer coefficient, A_f represents the wetted surface area of the channel, and $\dot{m}$ is the coolant mass flow rate. Lateral conduction resistance from the bulk of the cell and fin to this surface is assumed to be comparatively negligible.

Some of the attributes of this model are tuned to the Volt-type prismatic liquid-cooled arrangement where the "fin" (a thin aluminum sheet with micro-channel grooves, covering the area of the cell) is thermally more significant than the fluid it contains. For other designs with negligible fin mass, body f (see Fig. 4.6) becomes pure fluid, and it may be possible to use simplified convection formulas. In such cases (e.g., air-cooled cells), users can easily adapt this template in Simplorer, to create and store a customized unit model. For pack-level models that include CFD (co-simulation) or ROM units, the following quantities are reported:

1. Pack-level thermal uniformity metric

$$M(t) = \max_j \left\{ \max_k \left| T_{c_j} - T_{c_k} \right| \right\} \qquad (4.8)$$

where j and k are cell indices. A history plot of M shall be presented and its temporal average value reported.

2. Coolant temperature at the pack manifold outlet

$$\overline{T_o} = \sum_j \dot{m}_j T_{o_j} / \sum_j \dot{m}_j \qquad (4.9)$$

where j is the cell index. A history plot of $\overline{T_o}$ shall be presented and its temporal average value reported.

3. Worst-case time and location

$$T_{max} = \max_j \left[\max_k \left\{ T_{c_j} (t_k) \right\} \right] \qquad (4.10)$$

where j is the cell index and k is the time-step index. The script shall report j, t_k, and T_{max}.

4. Tabular listing in the report, for each unit in the pack, of:
    (a) Final cell SOC and cell voltage.
    (b) Temporal min, max, and avg. of cell spatial-average temperature.
    (c) Temporal min, max, and avg. of coolant bulk-temperature gain through channel $(T_o - T_i)$.
    (d) Temporal peak rate and total time-integral of heat generated by cell.

4.3.3 Reduced Order Models

As depicted previously in the ANSYS Battery Design Tool (ABDT), [14] is the framework or "umbrella" over all capabilities, including the graphical UI layer that automates and customizes battery simulation workflow, leveraging and enhancing ANSYS's Workbench and other existing commercial products.

Field simulation and system simulation, along with innovative methods to blend them together, are the key technologies that we have integrated. Co-simulation

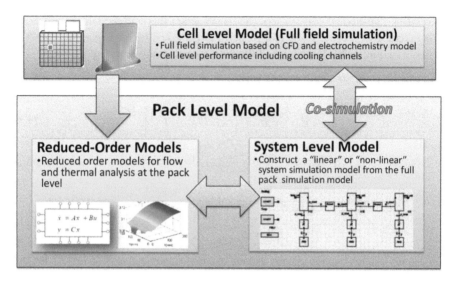

Fig. 4.7 CAEBAT ROM workflow

simply means using both methods together via run-time coupling between solvers. Based on domain decomposition, field simulation can be used for a small portion of the pack to limit the computational cost.

The team pursued several ROM approaches involving different choices for the "unit" to be reduced and the strategy for coupling the various physics. Figure 4.7 shows the general ROM strategy, which allows a wide range of options so that users can trade off computational expense versus accuracy.

4.3.3.1 ROM Approaches

There are two approaches to ROM development that we have considered in CAEBAT, one that we call black-box by treating the underlying physics and its inner states as black-box while only caring about its input and output responses. We call the other method direct in which the state-space matrices or its variant are directly provided by the solver (Fluent). Here is a comparison summary between black-box and direct methods:

Black-box ROM:

- Direct access to the true dynamical system is not available.
- System identification of the dynamical system is based on a set of persistent excitations and their corresponding responses used as training data.
- The generated ROM will only describe the relationship between the inputs and outputs of the dynamical system:

  - No knowledge of what happens inside the dynamical system.

- If the ROM is in state-space form, then the states do not have physical meaning.
- No guarantee that the ROM can predict the behavior of the dynamical system for excitations not included in the training data.

Direct ROM or ROM with direct access to the matrices of the dynamical system provided by the CFD solver (Fluent):

- If in state-space form, the reduced states can reach the true states of underlying dynamical system using appropriate projection matrices.
- Almost all of the information about the true dynamical system can be recovered.
- Example: Krylov projection-based moment matching ROM.

4.3.3.2 ROM in System Modeling Approach

A summary of the ROM approach in system modeling has as follows:

- Run pack network flow to calculate pipe velocities, pressures, and mass flow rates of micro-channels.

  - Create Lookup table ROM from the mass flow rate samples.

- Create electrical circuit model (ECM), after using appropriate parameter extraction methods, to calculate generated cell voltage and (uniform) heat source.
- Create thermal ROM for the unit cell models.
- Couple the ECM with thermal ROM to create an electro-thermal ROM unit in a multi-physics tool (Simplorer).

With the direct ROM approach, we have developed a thermal ROM tool that is able to provide accurate approximation to the 3D temperature field based on Krylov projection method [15] which will be discussed in Sect. 4.4.2.6.

With the black-box approach, we segregated the nonlinear physics, i.e., the electrochemical behavior of the battery, by using an ECM representation, from the thermal part which is approximated by a linear time-invariant (LTI) state-space ROM when the mass flow rate is fixed and a Linear Parameter Varying (LPV) ROM when mass flow rate is varying using a uniform heat source as input to the system. These methods will be discussed in separate sections in the report.

4.3.3.3 Development and Application of LTI ROM Automation

The LTI automation tool automatically generates a set of step response training data validated against ANSYS Fluent benchmark and meeting a prescribed ROM quality. The LTI automation tool also generates a Simplorer Model Library (SML) file holding the thermal ROM model imported into Simplorer and placed into the schematic prepared for system modeling.

Before we get into a summary of the automation tool, we present the application of system modeling with LTI ROM on a production-level GM 24-cell battery module with foams placed in between each two-cell battery unit within the module.

4.3.3.4 Application of LTI Thermal ROM to GM Battery Module

The ECM R-C fitting and parameter extraction was performed by GM using standard HPPC measurements. A set of thermal step response training data was generated. A Simplorer system model of the ECM combined with the thermal LTI ROM was used to approximate the electrical and thermal behavior of the battery module. The results were also directly compared with measurements when possible. Figures 4.8, 4.9 and 4.10 show the comparison between the system model with LTI thermal ROM against Fluent and measurement results.

The results of generating the voltage and thermal behavior of the battery compared with both CFD and measurements using the CAEBAT system modeling with LTI ROM approach applied to the GM production-level 24-cell battery module are in good agreement.

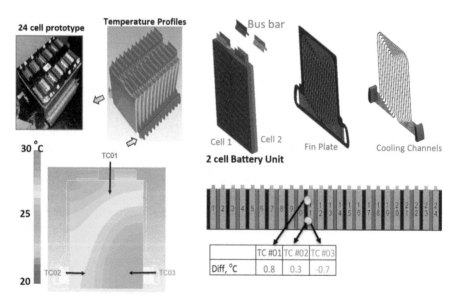

Fig. 4.8 A 24-cell battery module used for system modeling with LTI ROM validation

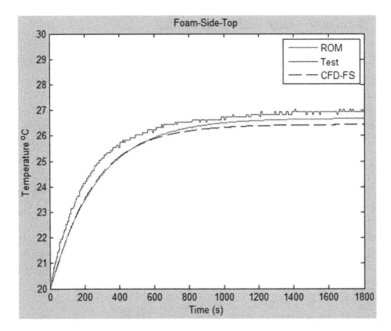

Fig. 4.9 Volume averaged temperature comparison: foam-side top

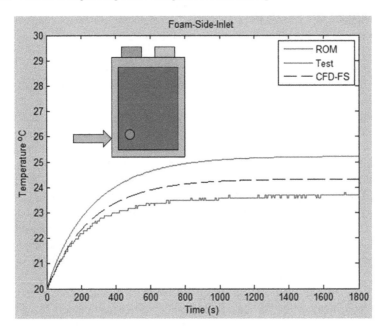

Fig. 4.10 Volume averaged temperature comparison: foam-side inlet

4.3.4 Thermal Abuse/Runaway Model

Lithium-ion batteries are susceptible to extreme electrical, mechanical, and thermal loads outside the normal operating conditions, compromising the safety of rechargeable energy system and its components. The control system typically prevents the battery cell from charging the cell beyond a specified voltage limit. In case of battery overcharging the battery could overheat and severe exothermic reactions might occur. During an external short, abnormally large current will pass through the cell causing resistive heating of battery and on-setting rapid decomposition reactions. Internal cell faults could occur due to manufacturing defects in raw materials used in the lithium-ion cells. These contaminants can trigger short current within the cell causing local heating and decomposition of electrode, eventually leading to thermal runaway even during normal charge operation.

Simulation can be invaluable to study the Li-ion electro-thermal behavior under abuse with or without an effective battery cooling system. Faults can be introduced in the simulation system model, and the measurements obtained from the previous testing will serve as validation of the simulation results. One can utilize the cell-level testing conducted using various abuse conditions and combine with reaction engineering models to develop a 3D thermal runaway simulation model implemented in ANSYS Fluent. GM team developed design and technical requirements during model development, so the features provided in the final tool can be immediately applied by product development engineers in the industry. These CAE tools can be used to assess and improve the abuse tolerance of the battery packs with and without active cooling.

Hatchard et al. [16] developed a heat transfer model that considers Joule heating of the cell during thermal runaway and empirical correlations for the parameters dependent on temperature. A similar model that includes heat generated from chemical reactions was proposed by Kim et al. [17] for a more general cell shape. The model was further refined and demonstrated the three-dimensional effect of chemical reactions and associated heat flow in abuse conditions such as internal short circuit events. Spotnitz et al. [18] extended Hatchard's model to three dimensions and demonstrated the heat dissipation properties in cells of different sizes. In addition, Smith et al. [19] developed a model that can address cell-to-cell interaction and the propagation of thermal runaway in a battery pack. Other efforts [20] toward understanding the behavior of the lithium-ion cell during an internal short circuit include the distributions of the various abuse reactions as well as temperature within the cell to better understand the phenomenon of internal short circuit in light of cell safety.

Physics-based abuse models are more flexible in evaluating safety of various cell designs and chemistries. They use abuse kinetic reaction parameters like activation energies, heats of reaction for anode, cathode, separator, and electrolyte individually and then combine them to evaluate overall cell-level heat generation. This requires knowledge of cell composition as well as material properties that are not readily available to OEMs from the cell manufactures due to intellectual property protection

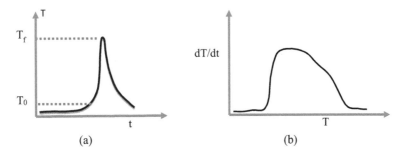

Fig. 4.11 Typical ARC temperature measurement and self-heating rate

as well as lack of experimental data. The semiempirical approach proposed in this project uses the heat generation rate data obtained from accelerated rate calorimeter (ARC) and uses that information in a 3D thermal cascading model as heat source assigned to battery cell components. Thus, the thermal abuse assessment will be limited to the selected cell design and chemistry and cannot be generalized. The kinetic reaction parameters are numerically calibrated based on ARC data providing a practical advantage when cell design composition is unavailable to the user. In a typical ARC experiment, temperature vs time data (T vs t) on surface of battery cell is collected from onset temperature T_0 to a final temperature T_f. Once the abuse reaction is complete, the heat generation within battery due to decomposition reactions is assumed to be zero and the battery cools down. Exact time at which onset of abuse reactions take place to a point of no return can be determined by monitoring the rate of change of temperature, i.e., $\frac{dT}{dt}$. Also, the heat generation due to decomposition reaction is assumed to be negligible before the onset temperature T_0 is reached. Figure 4.11a shows a typical ARC temperature data not related to any particular abuse mode and cell design. Figure 4.11b illustrates the rate of temperature change plotted against cell temperature.

The reaction kinetic model implemented in CAEBAT project is based on work by MacNeil and Dahn [21]. In a thermal induced reaction in solids, the reactants are converted to products by a rate

$$\frac{d\alpha}{dt} = -k(T)f(\alpha) \tag{4.11}$$

where the rate constant

$$k(T) = \gamma e^{-\frac{E_A}{T}} \tag{4.12}$$

Further, α is the fraction of reactants available at time t, $f(\alpha)$ is the reaction model to be derived for a particular abuse test performed in the ARC, and γ, E_A are Arrhenius parameters (the pre-exponential factor and the frequency factor respectively).

A universal reaction model has been proposed that assumes the mathematical form

$$f(\alpha) = \alpha^m (1-\alpha)^n (-\ln(1-\alpha))^p \tag{4.13}$$

In many cases, setting the coefficient $p = 0$ seemed adequate while calibrating the model.

In an ARC experiment, the self-heating rate of the battery cell may be equated to

$$\frac{dT}{dt} = -\frac{d\alpha}{dt} \cdot \frac{H}{C_{total}} = -\left(\frac{d\alpha}{dt}\right) \Delta T \tag{4.14}$$

where H is the total heat energy released from the battery cell due to the abuse reaction, C_{total} is the total heat capacity of the reactants, and $\Delta T = T_f - T_0$ is the difference between onset and final temperatures during reaction. Upon consolidating the above equations, we get

$$\frac{dT}{dt} = (\Delta T)\, \gamma e^{-\frac{E_A}{T}} \left(\frac{T_f - T}{\Delta T}\right)^m \left(\frac{T - T_0}{\Delta T}\right)^n \tag{4.15}$$

The ARC experimental data in the format shown in Fig. 4.11b can be used to calibrate the reaction model parameters EA, γ, m, and n using any standard statistical software techniques and tools available to the user. Once all the relevant model parameters are obtained, the heat source to be included for battery cell in 3D Fluent model is

$$Q(T, t) = mC_p \frac{dT}{dt} \tag{4.16}$$

where m is mass of the cell and C_p is specific heat capacity of the cell. The implementation of this proposed heat source model in Fluent is not straightforward and requires user-defined functions (UDF). For instance, when battery cells in a pack model are assigned a custom heat source and the reaction is complete, the temperature of cells decrease below T_f, and the model might start the reaction again. In other words, the fraction of reactants, α, will suddenly increase which is physically not possible. Thus, modeling the heat source as a function of temperature alone is not adequate; a second variable α must be tracked as well to ensure the fraction of reactants available will go from 1.0 to 0.0 during the reaction only once during a simulation. Thus, α and $\dot{Q}$ are solved using these coupled equations:

$$\alpha = 1 - \frac{\int_0^t \dot{Q} dt}{mC_p \Delta T} \tag{4.17}$$

$$\dot{Q} = mC_p (\Delta T)\, \gamma e^{-\frac{E_A}{T}} (\alpha)^m (1-\alpha)^n \tag{4.18}$$

Two UDF have been defined as adjust and source functions in Fluent. Two user-defined-memories (UDM) are defined to save the heat generate rate $\dot{Q}$ and α. A scheme file was created to input the abuse kinetic reaction model parameters into a Fluent UI panel. A more complex model can override the scheme file through use of UDF and recompilation. UDF are based on C-programming language, and any custom numerical implantation and flags can be defined by the user. To initiate the thermal runaway in battery cells, "Patch" function in Fluent can be used to set the temperature of cells of interest above T_0 which will initiate the kinetic reaction. If the initial temperatures across the battery pack are below T_0, no cascading might occur without a secondary heat source as α remains at 1.0 and $\dot{Q}$ remains zero. Alternatively, α cannot be zero to initiate the thermal runaway. A screenshot of scheme file implementation in GUI is shown in Fig. 4.12. Modeling of vented gases, combustion, and flames are not modeled in this framework. Subsequent convective transfer of heat by vented gases as well as mass balance is also not addressed. The specific abuse conditions and processes involving nail penetration, overcharge, or an external short require solving electrical and thermal responses simultaneously.

The runtime for a typical cell-level thermal abuse model was within a few seconds, since only the energy equation was solved without any flow calculations. The abuse case was run on a cell mesh that previously was used to model

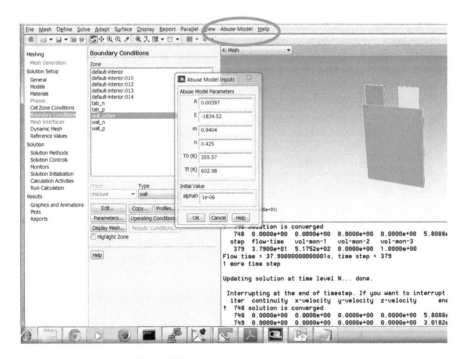

Fig. 4.12 Scheme file in Fluent GUI

electrochemical behavior upon tuning of the MSMD model. Typical cell level results are shown in Fig. 4.13.

Next a small battery module with three cells in series and later a module with ten cells were used to verify that the framework can be implemented beyond a single cell. One of the cells in the module was given a large initial temperature to onset the reactions. However, there was no thermal cascading to the remaining cells due to large air gaps between the cells in the model. Then the air gap was meshed and modeled for heat conduction in ten-cell module. The resulting temperature plots are shown in Figs. 4.14 and 4.15, respectively. Heat transfer through cell tabs was minimal in this case. Heat transfer coefficients were used to model the boundary condition for the face of the cells.

The effect of flow of air on the thermal cascading propensity can be studied using this model. Initially, a flow rate of 0.0001 m/s for air was used to simulate no-flow condition. In the ten-cell battery module, Cell 5 was triggered for thermal runaway. Then the effect of thermal conductivity of the medium in the gap between cells and flow rate on how quickly cascading occurs can be studied. Two levels of flow rate, 0.001 m/s and 0.2 m/s, and two levels of thermal conductivity, 0.0242 W/m/K for air and 2.42 W/m/K for an arbitrary conduction medium between cells, were considered. The results in terms temperature of various cells in the module are shown in Fig. 4.16.

In summary, the UDF structure and implementation in Fluent was simple to use even for a beginner with little training. Extending the framework to large-size module and pack scenarios that include flow and radiation should be straightforward. In realistic packs, frames and other structural members around the battery cell might provide more conduction paths for thermal cascading to occur.

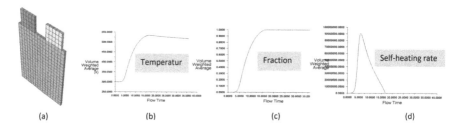

Fig. 4.13 (a) FE mesh of cell, (b) temperature, (c) fraction of reactants, (d) self-heating rate

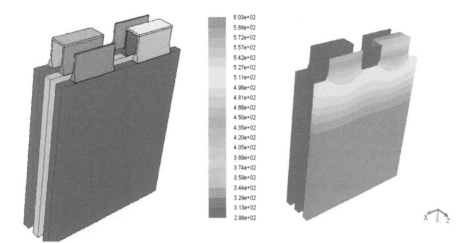

Fig. 4.14 1P-3S cell configuration temperature contours

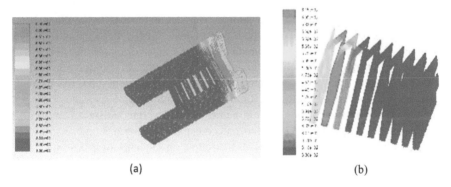

Fig. 4.15 1P-10S cell configuration temperature contours. (**a**) Air mesh, (**b**) battery cells

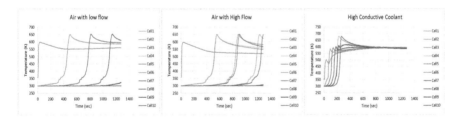

Fig. 4.16 Sensitivity analysis on thermal cascading

4.4 Battery Test and Validation

4.4.1 Cell Performance Test and Validation

4.4.1.1 V&V Process Overview

The accuracy and usefulness of the MSMD cell-level models were demonstrated through rigorous verification and validation processes during CAEBAT project. The proposed framework for verification and validation of battery multi-physics models is based on the process developed with GM-NISS collaboration [22, 23] as well as the ASME PTC60 guide [24]. Verification considered issues related to software quality, programming errors, numerical error estimation, and solution convergence. Thus, the verification work preceded validation and did not require any physical test data.

The proposed requirements for verification of battery physics mainly consist of code verification and solution verification. In code verification, we check for mathematical correctness of equations, understand specific algorithms used to implement the method, and perform software quality assurance feedback via documentation, version control, how code architecture is developed, and benchmark using analytical solutions for special cases or simplified PDEs. There are additional consistency tests performed for unit system, charge and mass conservation. In solution verification, we study convergence of model output with mesh size, element type, and meshing scheme; understand limitations of the solver for input C-rate, parameter ranges; identify issues of singularities, discontinuities; as well as standardize solver solution controls such as relaxation parameters and convergence criteria.

The model validation work involves the following sequence of activities:

1. Specify model inputs and parameters with associated uncertainties and ranges.
2. Determine the evaluation criteria or the response quantities of interest.
3. Collect data through design of experiments: computer and physical.
4. Approximate computer model outputs.
5. Compare model output to the test data.
6. Feed validation lessons into the model for future predictions.

The input material properties and cell-specific parameters are characterized along with associated uncertainties. The collection of input parameters is carried out through literature review, supplier data, and new measurements. Prior sensitivity analysis or screening will help avoid measuring parameters that are not relevant. Typical quantities that are compared against test data include voltage versus depth of discharge, dissipative heat, and skin temperature profiles. Since there are many variable parameters that can affect the battery performance, only a few important variables like charge/discharge rate and profile and ambient conditions are varied in physical experiments. Bias or model form error is computed as a function of model input parameters so that the final model prediction can be adjusted with bias value for any future untested conditions. GM generated the test database for the physical validation of the cell-level electrical and thermal performance.

4.4.1.2 Cell Performance Test Setup and Data

Temperature and voltage responses were chosen as evaluation criteria. These response quantities of interest predicted from CAE model will be compared to the corresponding experimental results to judge the model accuracy. Voltage is measured across the terminals of the cell, while temperature is monitored at seven locations of the prismatic cell chosen for testing in the project. The cell test setup and sensor locations are shown in Fig. 4.17. The cell was placed in a thermal chamber where ambient conditions with different temperatures can be set.

The cell is connected to a cycler so that various C-rates current can be drawn while maintaining voltage cutoff limit specified for the supplier. After the cell reaches minimum voltage value, cell will be maintained at rest (zero current) and allowed to cool down. Table 4.3 shows the design of physical experiments matrix with various ambient temperatures and applied currents. The convective heat transfer film coefficient (h_{film}) between surface of battery and air stream in the cycling chamber was derived using battery cooling curves. Cables attached to the tabs might be taking the heat away, and hence the estimate of convection heat transfer coefficient for tabs has some uncertainty.

Figure 4.18 shows the various thermal zones the cell experiences during cycling and rest. Zone III or the rest period is used to compute the heat transfer coefficient.

In zone I, the cell heat generates more heat than it dissipates, and hence the cell temperature rises due to accumulation of heat. In steady-state condition in zone II, the heat dissipation rate is the same as the heat generation rate, and hence the heat accumulation rate is zero. In zone III, the cell does not produce any heat, and hence

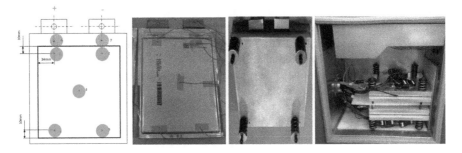

Fig. 4.17 Cell test setup for thermocouples

Table 4.3 Physical experiment design matrix

Case	T_{amb} (K)	Current (A)	h_{film} (W/m²/K)
1	253	12.06	25
2	253	7.46	27
3	275	45.08	26
4	275	14.8	25
5	275	7.44	18
6	297	60.19	25
7	297	45.08	25
8	297	15.08	23
9	297	7.54	23
10	312	60.19	23
11	312	45.7	23
12	312	15.08	20
13	312	7.47	16

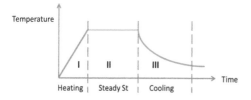 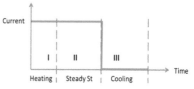

Fig. 4.18 Various thermal conditions of cell

any heat loss is due to the convective cooling of the cell in the convective thermal chamber. Mathematically, the process in zone II can be described as

$$mC_p \frac{dT}{dt} = -h_{film} A \left(T - T_\infty \right) \tag{4.19}$$

where m is mass of the cell, T is temperature, C_p is the specific heat capacity of the cell, and A is total area on cell exposed to the ambient temperature T_∞. The above equation is solved and matched to the test cooling curve (temperature vs time) by adjusting the heat transfer coefficient h_{film}. We also adjusted similar coefficient in ANSYS Fluent model and matched it to the test data and found that the results are very close to the lumped approach of tuning h_{film}.

4.4.1.3 Model Verification and Input Uncertainty

For the code and solution verification of MSMD model implemented in Fluent, we studied the solution control settings that resulted in submitted job to fail. Thus, feasible ranges for relaxation factors for smoothing and solver convergences have been identified. The under-relaxation factor for current should be in the range of 0.4–1.0 at any C-rate. Beyond this range, convergence was not obtained with the solution

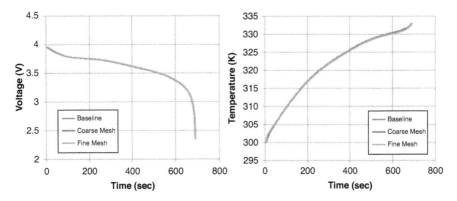

Fig. 4.19 Mesh convergence for NTGK model

overshooting to maximum limits specified in the model. The relaxation factors for φ_+ and φ_- may be left at default value of 1.0. We noticed that the model converged even at high C-rates (20° C) while using model parameters in UDF, lookup table format, or polynomial functions. The cell-level battery model was run using three levels of finite element mesh coarseness, and we found good grid/mesh convergence as shown in Fig. 4.19. The mathematical equations implemented in NTGK, ECM, and P2D models were verified against literature, boundary conditions verified and recorded as part of discussion with ANSYS code developers.

Before the 3D model could be run, there are some model inputs whose values are not well-known or measured with certainty. For instance, the through-plane thermal conductivity of the battery cell in the model affects the temperature distribution on the face of the cell in addition to the current density distribution due to cell aspect ratio and terminal tab configuration. Figure 4.20 shows the effect of in-plane thermal conductivity on battery thermal distribution. The measured scatter in temperatures at seven locations was used to adjust the thermal conductivity in the 3D Fluent model.

The through plane thermal conductivity measurements were obtained separately by the cell supplier. Similarly, literature data, repeated measurements, and other estimates were used to derive the values for additional model input parameters like specific heat capacity, mass, nominal discharge capacity, current collector electrical resistance, etc.

4.4.1.4 NTGK Simulation Results

The primary usage of NTGK model is for predicting the thermal behavior of battery cells. The model parameters U and Y must be calibrated using cell voltage data obtained under various C-rates, states of charge (SOC), and ambient temperature. We found eighth order polynomial functions are needed to capture U & Y as a function of SOC. Since the voltage data already includes the effect of cell

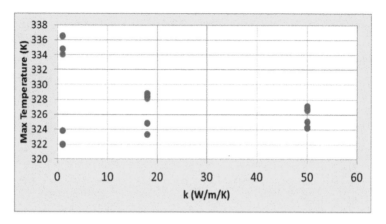

Fig. 4.20 In-plane thermal conductivity sensitivity

temperature, U & Y should not be modeled as function of ambient temperature additionally. Thus, the parameters C_1, C_2 are not used in the model prediction. Only cell temperatures between test and simulation are compared for this model as the voltage responses are already used for model parameter estimation and hence cannot be used again for model validation. Figure 4.21 shows the comparison between model and test average temperature (from five locations) responses.

The ambient air temperatures specified in the DOE matrix are not exactly maintained constant throughout the experiment as is evident from thermocouple data placed near the surface of the cell but not attached to it. One might observe an increase of 1.5 degrees during cycling which might be contributing to heating up of cell through convection. Thus, any discrepancy from experimental results is attributed to input uncertainties to the model that must be properly accounted for before making any conclusion regarding the predictive accuracy of the model. A statistical analysis of model error at various DOD, ambient temperature, and C-rates as summarized in Fig. 4.22 through analysis of variance (ANOVA) plot indicates that the model error (Test – Simulation) is larger at lower ambient temperatures and higher depths of discharge. Similar trends were observed for temperature prediction errors at seven different locations.

4.4.1.5 ECM Simulation Results

The equivalent circuit model implemented for the purpose of demonstration of model validation is shown in Fig. 4.23. In this six-parameter or 6P model, the resistances, capacitors, and open circuit voltage parameters are calibrated using HPPC test voltage data at various SOCs and ambient temperature. The pulse current is usually at constant value of 5C or 10C discharge rate for 10 s duration followed by rest for 40 s before applying charge pulse of similar C-rate current. After the pulse test, the cell is discharge at 1C rate until SOC decreases by 5% or 10%. Unlike

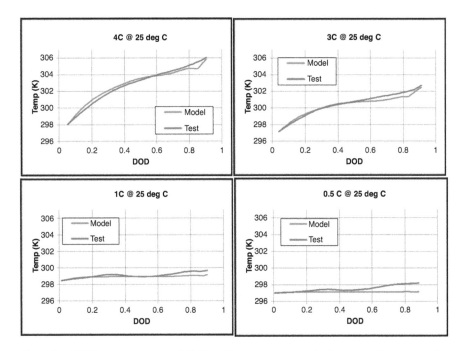

Fig. 4.21 NTGK model vs test at 25° C ambient temperature. Similar comparisons were also performed across a wide temperature range of −20 to +45 °C [25]

NTGK model where the cell temperatures change during cycling, they remained nearly constant in the HPPC test. Thus, the estimated model parameters strongly depend on ambient temperature and SOC.

Even though the charge and discharge pulses result in different set of parameters, the validation cases used in this study correspond to discharge only. Figure 4.24a shows the comparison between model and test average temperature (from five locations) responses at 0° C. Similar comparisons were made at different ambient temperatures [25]. Figure 4.24b shows the voltage response from ECM model at selected temperatures.

Since the chamber air temperature varied during the experiment, the discrepancy between model and test temperatures at lower C-rates might be attributed to the measurement error. Also, independent internal data at GM on the tested cells as well as similar cells show that ECM resistances derived from 5C pulse are much lower than those from 3C or 1C pulses. The effect of C-rate is particularly dominant at lower temperatures and higher depths of discharge. Thus, the resistance values computed from the HPPC test at 0 and − 20 degree ambient conditions might be underestimated. Similar validation was performed at different temperatures using the P2D models [25].

Table 4.4 summarizes the calibration and validation data used for various cell level sub-models. It is clear from the results that the accuracy of temperature

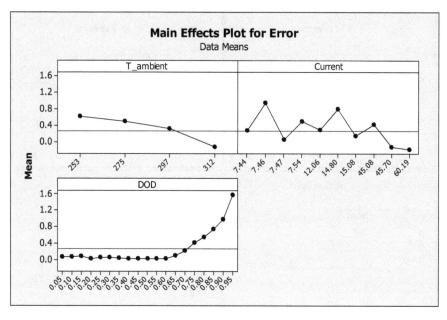

Fig. 4.22 ANOVA plot for model error (Ave. temperature)

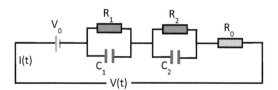

Fig. 4.23 6P Equivalent circuit

predictions depends on how well the voltage responses match/meet simulation and physical experiment.

Apart from model form selection or a particular sub-cell model, the parameter estimation and corresponding confidence intervals are key to model validation and when comparing to test data. Even though average surface temperatures from model matched well with average data from five thermocouple data, the actual temperature distribution and gradient developed were not captured well by the model. The in-plane thermal conductivity boundary condition as well the tab connections in the laboratory experiment may not have been accounted properly in the simulation. All the three cell sub-models provide a great framework and platform for simulating coupled electrical and thermal physics for the battery thermal management needs.

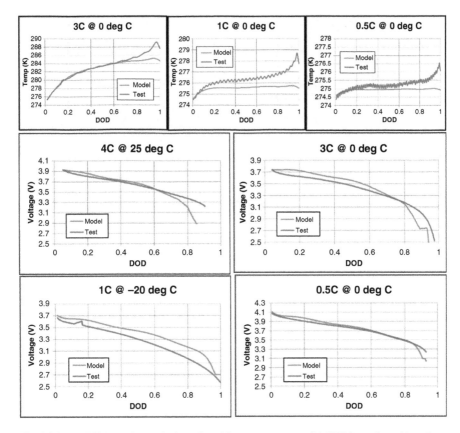

Fig. 4.24 (**a**) ECM results at 0 deg. C ambient temperature. (**b**) ECM results with voltage comparison

Table 4.4 Cell sub-level models

Model	Calibration data	Validation data
NTGK	Voltage curves from constant discharge current at various ambient temperatures	Temperature measurements across face of the cell
ECM	Voltage curves from HPPC test at various ambient temperatures	Terminal voltage and temperature measurements across face of the cell
P2D	Voltage curves from HPPC test at various ambient temperatures	Terminal voltage and temperature measurements across face of the cell

4.4.2 Module/Pack-Level Validation

The field simulation approach directly solves flow, energy, and electrochemistry governing equations for the entire model. It provides a solution with the highest spatial resolution, and also it is the most time-consuming and resource-intense approach. With the current computational resources, the field simulation is required

Fig. 4.25 24-cell validation module

to speed up the computational efficiency to handle transient problems for various vehicle driving cycles.

4.4.2.1 Field Simulation Test Case

To demonstrate the field simulation capability, instead of using production packs, the GM team has decided to utilize a 24-cell validation battery module as shown in Figs. 4.25. The module was built by GM to gain understanding on battery pack behaviors in the early stage of battery pack development. Since it is a learning prototype module, the properties of the module are well understood. This learning prototype module was designed to investigate the cooling performance of the liquid cooling design in the module under various driving cycles. The detailed thermal data is available for this learning prototype module, and this prototype module is an ideal choice for the validation of CAEBAT pack-level tools.

The 24-cell validation module consists of 12 repeated units. Each unit has two pouch-type battery cells and a liquid cooling fin with micro-channels between the battery cells. Each unit also has connectors/bus bars and two layers of foam which provide insulation and isolate the unit from its neighboring units. A schematic of a unit is shown in Fig. 4.26. Electrically, the two battery cells in a unit are connected in parallel, and then the individual units are connected in series to form a 2P12S (2 parallel, 12 series) configuration.

As shown in Fig. 4.26, a liquid cooling fin is placed between the two battery cells. A schematic of liquid fin is shown in Fig. 4.27. Similar to a typical cooling fin, the liquid fin is designed to remove the heat generated by the adjacent battery cells, and the heat is dissipated to the ambient environment. Unlike a typical fin, the liquid fin not only utilizes its high conductive solid plate but also the coolant inside the fin to convect the heat generated from the battery cells. Liquid coolant is

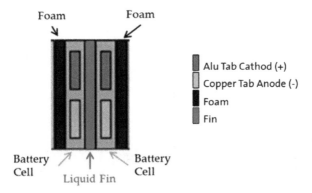

Alu Tab Cathod (+)
Copper Tab Anode (-)
Foam
Fin

Fig. 4.26 Unit schematic with a cooling fin and two foams (top view)

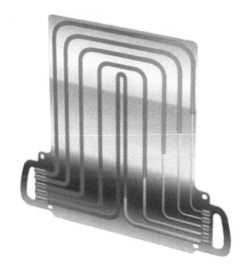

Fig. 4.27 Schematics of a liquid fin design

forced to flow through several micro-channels in the liquid fin. The current liquid fin design enhances the cooling performance by a convective heat transfer. The micro-channels are designed to keep the battery cells in an optimum temperature range and furthermore to improve temperature uniformity within the battery cells. The liquid fin is pressed against both battery cells to ensure good contact between the fin and neighboring cells.

The coolant is fed into the inlet manifold and then enters into the micro-channels in the liquid fin. In the 24-cell module cooling system, the 12 liquid fins are connected in parallel by the inlet/outlet manifolds. The cooling system design can be viewed as 12 flow circuits connected in parallel between the inlet and the outlet manifolds. The flow area of the inlet manifold is relatively large compared with the flow area of the micro-channels to provide relatively uniform flow distributions for

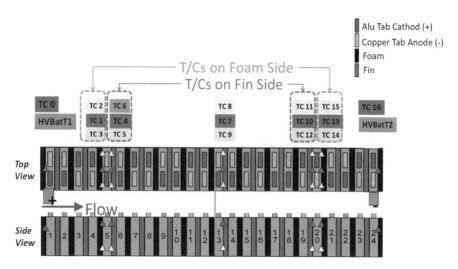

Fig. 4.28 Locations of temperature measurement in the 24-cell validation module

12 liquid fins in the module. Insulation foams are placed on both ends of each unit. The foam layers accommodate the cell expansion and contraction during charge and discharge events and also provide thermal isolation between the cell units. The total coolant flow rate for the module is maintained at a constant mass flow rate.

For this learning prototype module, GM instrumented many thermocouples inside of the module to evaluate the cooling performance of the current liquid cooling design. Figure 4.28 shows the temperature measurement locations within the 24-cell module. T-type thermocouples are installed at 17 locations to monitor the module temperature. As show in Fig. 4.28, both sides of Cells 5 and 20 are monitored by six thermocouples: three on the battery cell side facing the liquid fin (fin side) and the other three on the battery side facing the insulation foam (foam side). The arrangement of thermocouples is shown in Fig. 4.29. The three locations are chosen with the intention to capture colder (near the coolant inlet) and hotter spots (near the coolant outlet and tabs) on a cell surface. For Cell 13, the fin side is monitored with three thermocouples similar to those on Cells 5 and 20. The foam side of Cell 13 is monitored by a temperature board with 30 thermistors (5×6 matrix) which is able to measure the spatial distribution of the entire battery surface temperature.

First, a 170A current pulse was used for the validation of field simulation. The battery pack/module was then subjected to a repeated current loading of 170A discharge for 1 s followed by 170A charge for 1 s until the end of the test. During this pulse test, the battery module SOC stayed constant from an initial 50% SOC. The alternating charging/discharging current essentially generates heat at a roughly constant rate due to a constant SOC during the test. With cooling system running at a constant mass flow rate, the battery module temperature will continue to increase from its initial value and reach a steady-state value.

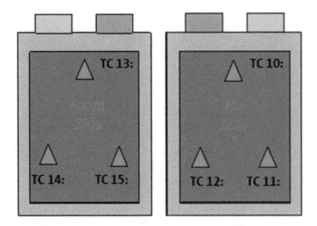

Fig. 4.29 Thermocouple locations on Cell 20

4.4.2.2   CFD Model Development

A Fluent CFD model was built to simulate the 24-cell module for the validation. The process involved building first a complete unit and then copying the first unit to construct the second, third, and up to the twelfth unit and translating the copied units to the correct positions. Finally, all units were connected to complete the entire module. Two battery cells including tabs and bus bars connecting tabs are meshed as one continuous solid zone. The solid portion of the liquid fin and manifolds are built into one continuous cell zone. The liquid portion of the liquid fin, which includes coolant, occupied zones in micro-channels, and both manifolds are specified as another continuous cell zone. The mesh interfaces between the solid and liquid zones of the liquid fin are conformal. Non-conformal interfaces are applied between battery cell and liquid fin. The total cell count of the 24-cell Fluent CFD model is about 23 million cells with predominately hexahedral cells.

Due to relatively small dimension of micro-channels and low coolant flow rate in the micro-channels, the Reynolds number is relatively small enough to justify the laminar flow in the micro-channels. The flow in the inlet and outlet manifolds can be turbulent, but the initial study showed that no significant difference was found between turbulent or laminar flows in the manifolds.

The battery sub-model used in the 24-cell module validation is equivalent circuit model with 6 parameters, as shown in Fig. 4.23. The empirical parameters are extracted from battery cell test data and are fit as a function of SOC. The same parameters are used for both charge and discharge cases. The size of the total computational model is about 23 million finite volumes, and the field simulation takes about 9.6 h for a 2000 second transient simulation with 64 CPUs of GM HPC system.

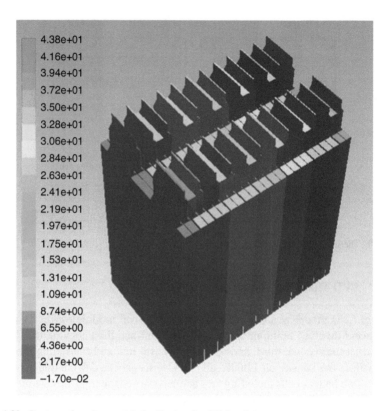

| 4.38e+01 |
| 4.16e+01 |
| 3.94e+01 |
| 3.72e+01 |
| 3.50e+01 |
| 3.28e+01 |
| 3.06e+01 |
| 2.84e+01 |
| 2.63e+01 |
| 2.41e+01 |
| 2.19e+01 |
| 1.97e+01 |
| 1.75e+01 |
| 1.53e+01 |
| 1.31e+01 |
| 1.09e+01 |
| 8.74e+00 |
| 6.55e+00 |
| 4.36e+00 |
| 2.17e+00 |
| −1.70e−02 |

Fig. 4.30 Contour plot of potential distribution for 170A pulsing

4.4.2.3 Electrical Validation

ANSYS implemented several new battery-related scalar variables into Fluent. These scalar fields are accessible to users via standard post-processing menu in Fluent. A contour plot of electrical potential for the 24-cell module is shown in Fig. 4.30. The potential distribution for 24-cell module is correctly simulated, and the potential distributions are consistent with the measurements.

In addition to the ability to plot battery scalar fields, ANSYS also added additional vector fields into Fluent. These battery-related vector fields are accessible to users via Fluent's standard post-processing menu. Figure 4.31 demonstrates a z-component vector plot of current density within the 24-cell module during the 170A pulsing simulation. The vector plot indicates the current flow direction within the module. The electrical connection of the bus bars is automated by the ABDT user interface during the model setup. A user only needs to specify the positive and negative terminals for the pack, and the ABDT automatically identifies the pack electrical configuration.

Fig. 4.31 Vector plot of electrical current density for 170A pulsing

Figure 4.32 shows comparison of voltage prediction vs measurement from the test with 170A pulsing. The voltages from simulation and measurement are in very good agreement. The voltage difference between simulation and test is less than 1%.

4.4.2.4   Thermal Validation

The battery pack total heat generation rate is nearly constant during 170A pulsing test at a constant SOC (50%). The temperature of 24-cell module reaches a steady-state value after an initial transient since the coolant flow rate and the coolant inlet temperature are kept constant during the test. Figure 4.33 shows the temperature contours after the temperature reached a steady state. As expected, the temperature is lower near the coolant inlet and higher near the tab area in the outlet side.

The battery surface temperature distribution is compared between the simulation and the measurement with the same color scale, as shown in Fig. 4.34. The simulation result agrees very well with the measurement. The black spots in the measurement indicate the data are missing.

Figure 4.35 shows the temperature contours from the simulation for the foam side of Cell 5 at the end of 170A pulsing. Figure 4.36 shows the comparison of the temperature-time histories at three thermocouple locations on the foam side of Cell 5. The table in Fig. 4.36 summarizes the difference between the simulation and the

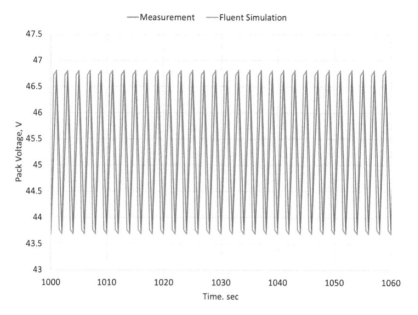

Fig. 4.32 Comparison of the module terminal voltage for 170A pulsing

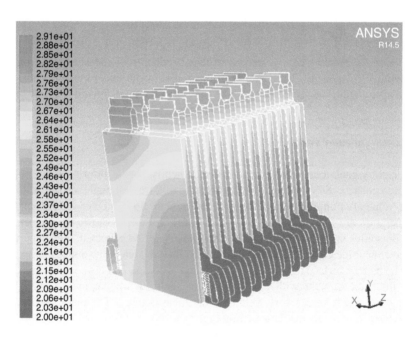

Fig. 4.33 Steady-state temperature contour during 170A pulsing

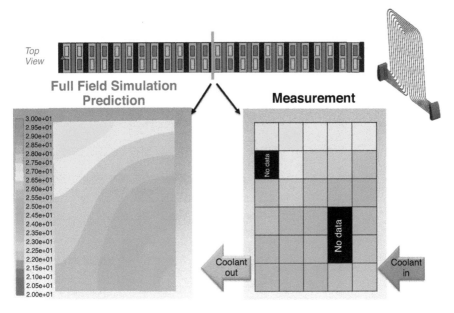

Fig. 4.34 Comparison of temperature distribution on the battery surface located in the middle of the module (Cell 13)

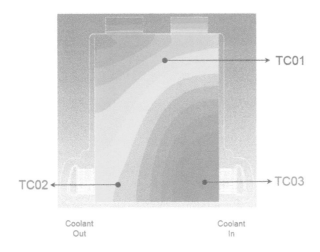

Fig. 4.35 Foam-side temperature contour in Cell 5

measurement. The agreement is excellent as the maximum discrepancy is less than 0.3 °C.

Figure 4.37 shows steady-state temperature contours from simulation on the fin side of Cell 5 at the end of 170A pulsing test. It shows the lowest temperature near the coolant inlet side while the hottest spot on the upper right side near the

Fig. 4.36 Three foam-side temperatures with time and the difference between the simulation and the test data in Cell 5

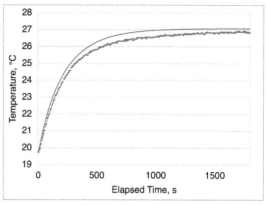

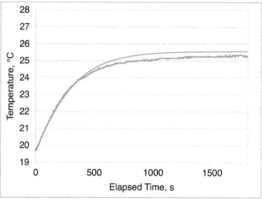

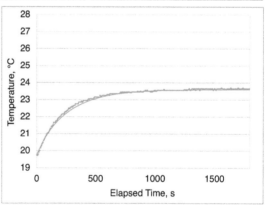

Simulation vs. Measurement @ 1800 s	TC01	TC02	TC03
ΔT, °C	0.3	0.3	−0.1

Fig. 4.37 Fin-side
temperature contour in Cell 5

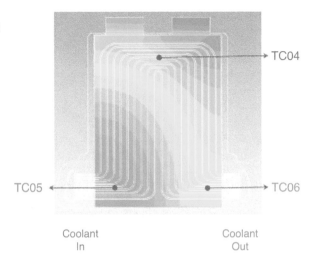

tabs. Figure 4.38 shows temperature-time histories of the three thermocouples on the foam side of Cell 5 between the measurement and the simulation (smooth line). The table in Fig. 4.38 summarizes the differences between the simulation and the measurement. The maximum difference is about 0.4 °C at TC04. The field simulation approach is successfully validated for the 24-cell prototype module. The good agreement between the simulation and the test results completes the pack-level validation for the field simulation approach.

4.4.2.5 System Simulation

The system simulation approach is useful and practical for battery pack transient simulations, for various vehicle drive cycles. ANSYS Simplorer was chosen as a base software for system simulation approach. Simplorer is a multi-domain, multi-physics simulation software. It integrates multiple system-based modeling techniques and modeling languages that can be used concurrently within the same schematic. This makes Simplorer an ideal system modeling tool for battery pack simulation. ANSYS developed dedicated scripts for battery pack modeling. The script, with friendly user interface, gathers user's inputs on battery pack configuration and automatically generates a customized battery pack simulation model. The user also has options to choose from the built-in battery unit template library or to create his own battery template library. In order to validate the system-level model without ROM approach, GM engineers constructed the system-level model from an automated ABDT user interface.

Fig. 4.38 Foam-side
temperature profiles in Cell 5

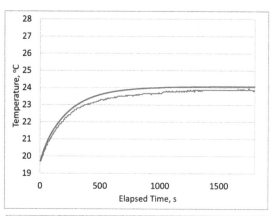

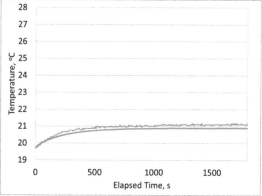

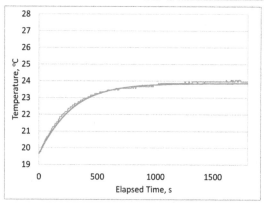

Simulation vs. Measurement @ 1800 s	TC04	TC05	TC06
ΔT, °C	0.4	-0.3	-0.1

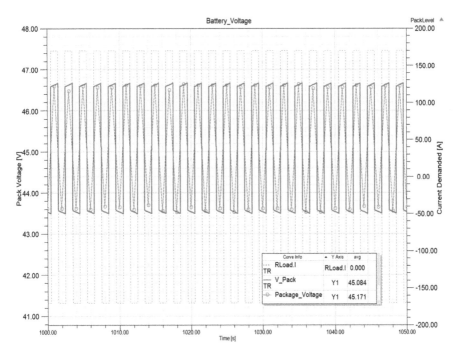

Fig. 4.39 Pack voltage of 24-cell module in 170A pulsing

Validation for 170A Pulsing Case

Figure 4.39 is snapshot of battery pack voltage for the 24-cell validation module during the170A pulsing test. In the figure, the blue line indicates simulation by the system model, while the red dashed line represents the data from testing. Because of the sampling rate used during the testing, the test data are only available on the peak values. The gray dashed line indicated the current loading with the axis on the right-hand side of the plot. As observed from the figure, the value of battery pack voltage estimated from the system approach model is in good agreement with the measurement from the testing.

Figure 4.40 is comparison of battery cell temperature predicted by system modeling simulation and the measurements during the 170A pulsing. The system modeling can capture the transient response very well. The system modeling slightly overpredicts the steady-state value, and the maximum difference between the simulation and the measurement is only 0.3° C.

The heat generation rate is critical in battery thermal simulation. Figure 4.41 shows the comparison of the heat generation between the simulation and the measurements based on the enthalpy difference between the coolant inlet and the outlet heat flows. The agreement for the total heat generation rate is satisfactory compared with the measured total heat rejection. The system modeling can capture the representative time-averaged value of heat generation. The system-level simula-

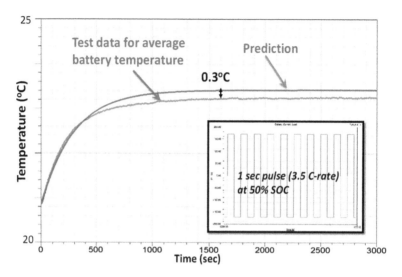

Fig. 4.40 Temperature of battery cell in 170A pulsing

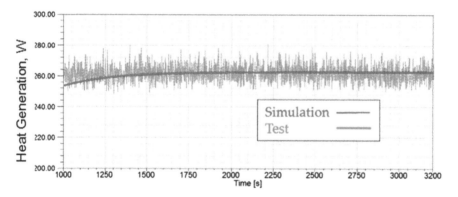

Fig. 4.41 Total heat generation of the validation module in 170A pulsing

tion demonstrated the total CPU runtime less than 1 min (Dell Z800 PC) compared to the field simulation that took 4 ~ 5 days with 64 processor HPC for the high-frequency charge/discharge pulse case.

Validation for US06 Driving Schedule

GM team has also developed a procedure to obtain empirical parameters from the HPPC test data that performs and predicts accurately the load voltage for various driving cycles and hence the total heat generation in the pack for US06 drive cycle. GM engineers also validated the system-level approach for a realistic US06 driving cycle. Figure 4.42 shows the battery pack voltage comparison between

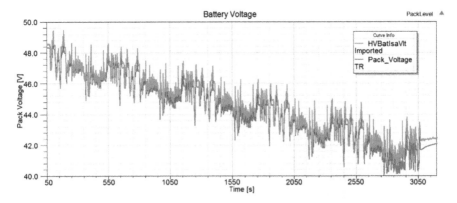

Fig. 4.42 Pack voltage of 24-cell module in US06

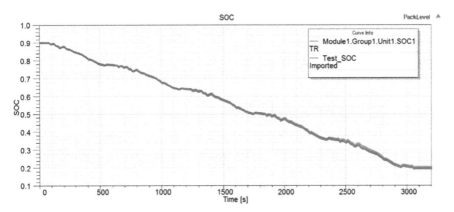

Fig. 4.43 State of charge for the 24-cell validation module in US06

the simulation (blue) and testing (red). The simulation and test data are in good agreement except toward the end of US06 drive cycle. Figure 4.43 shows the comparison of states of charge estimated from the system model (blue) and measured from testing (red) for 24-cell validation module in US06. The system simulation model demonstrated an accurate prediction of SOC under US06 drive cycle. The agreement for the pack total heat generation is satisfactory compared with the measured total heat rejection by the coolant mass flow rate and the coolant temperature difference between the inlet and the outlet as shown in Fig. 4.44. The validation of the system simulations for the 24-cell module is satisfactory, and the predicted temperatures were within 0.5 °C in comparison with the test data as shown in Fig. 4.44. Simulation of the five back-to-back US06 drive cycles for a total of 30 min driving cycle simulation took less than a few seconds in computational time with the system simulation. We demonstrated that the system simulation accurately characterizes the thermal behavior of the cells in the 24 cell module.

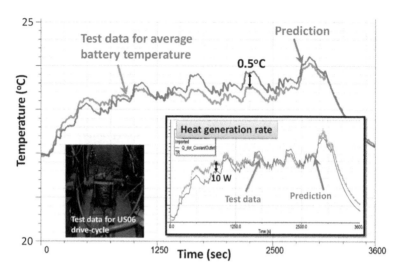

Fig. 4.44 Comparison of cell temperatures during US06 drive cycle and the heat generation comparison between the test and the prediction

4.4.2.6   ROM Verification and Validation

Reduced order modeling (ROM) is a mathematical technique that exploits redundancy in the original full order model by reducing orders of magnitude without sacrificing accuracy. There are essentially two types of ROMs: (1) direct and (2) indirect or black-box. In direct ROM, as the name suggests, the original set of governing equations applied to each finite volume/element (full order model) is first reduced using Krylov sub-space projection technique. The resulting ROM, orders of magnitude smaller compared to the original model, is then used in its place giving substantial computational speed up. In the black-box or indirect ROM, the original full order model is primarily used as a black-box, to yield the input and output required to characterize the system performance using well-established system identification techniques. No attempt is made to reduce the order of the original model at the governing equations level. Instead, as the name suggests, black-box or indirect ROM consists of an implicit encoding of the input-output behavior of the system. In this technique, care needs to be exercised in ensuring that the system (model) is sufficiently and persistently excited to yield response with maximum possible signal-to-noise ratio.

Test Setup for Validation

In this section, we will provide both verification and validation of the black-box LTI ROM's performance on a 24-cell 2P/12S liquid-cooled module as shown in Fig. 4.45.

Fig. 4.45 24-cell 2P/12S liquid-cooled module test setup

Fig. 4.46 24-cell battery module layout

Figure 4.46 shows the schematic of the module used in the test: the cells (light grey), the cooling fins (blue), insulating foam (black), and the positive and negative tabs (orange and gray). As seen in this figure, there is one cooling fin per two cells. The insulating foam prevents any significant heat transfer from one unit cell (two cells and one cooling fin) to the next. This, and the fact that cooling liquid flow rate through each cooling fin is almost identical, enables one to use only one unit-cell CFD/Thermal model. The model provides time-dependent step responses in the form of temperature at any desired location on the cell for a constant thermal excitation. Furthermore, the cooling flow rate to the module is also constant in time. This feature, constant flow rate, is absolutely essential for application of state-space linear time-invariant (LTI) black-box system identification modeling technique. In cases where the cooling flow rate does change with time, another mathematical modeling technique called LPV (linear parameter varying) is used. In this technique, local LTI models extracted at various discrete cooling flow rates suitably chosen between the minimum and maximum levels are assembled using interpolating techniques to yield a single LPV model that can now handle time-varying cooling flow rate.

As mentioned before, in order to characterize the thermal performance of this 24-cell battery module, a 2-cell and a cooling fin unit model are sufficient in this specific case instead of the entire 24-cell module model. Two key factors, (1) no significant heat transfer across the insulating foam occurs and (2) no significant cooling flow rate variation from one cooling fin to another, are responsible for this assumption. If the first assumption were not valid, the divide-and-conquer strategy applied wouldn't hold, and in that case, the full 24-cell CFD/Thermal model would

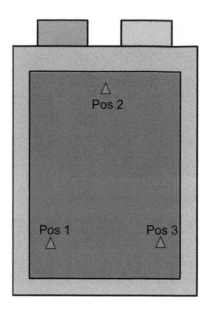

Fig. 4.47 Location of thermocouples on the cell surface. Three on the fin side and three on the other side (foam side) of the same cell

be needed to create the ROM. In the case of violation of the second assumption, LPV modeling technique will be required instead of just LTI.

In the physical test, the 24-cell battery module was subjected to an external load equivalent of a standard US06 drive cycle profile for a compact car. Based on an equivalent circuit model (ECM) of the cell's electrical behavior used in this module, the time-*averaged* rate of waste heat generation in the module is calculated as 227 Watts per US06 cycle. This is equivalent of approximately 9.4 watts of heat generation per cell. However, from the overall energy balance of the coolant, the module rejected a total of 267 Watts (difference in the coolant enthalpy between inlet and outlet to the module). The extra 40 Watts (267–227) can therefore be attributed to the heat generation in the tabs and the bus bar. This yields about 1.7 Watts of heat generation in each table A constant flow rate of the cooling liquid (50% water +50% ethylene glycol mixture) at 1.23 L/min and at 20C at the inlet of the module was applied to manage the cell temperatures to below 32C.

Three thermocouples are located on the fin side, and three on the foam side on Cells 5, 13, and 20 to measure the temperatures on the cells. Figure 4.47 shows these three locations on one side of the cell. There are three additional thermocouples at similar locations on the other side of the same cell. One of these sides is in contact with the cooling fin (fin side), and the other is in contact with the insulating foam (foam side).

LTI Model Verification Results

Before validating the black-box LTI ROM performance with the test data, we will verify its accuracy against the full-field CFD/Thermal simulations performed on the two-cell unit model as shown in Fig. 4.48.

This model has a total of six heat sources (inputs): 2 cells and 4 tabs and 16 (outputs), 6 temperatures at the thermocouple locations, 2 cell average temperatures, 4 tab average temperatures, and 4 tab-interconnect average temperatures. The two-cell unit CFD/Thermal model has approximately 1.9 million cells (1.1 million fluid cells and 0.8 million solid cells). In order to create the input-output data for state-space LTI ROM, the model is first run to steady state for flow only. Thereafter, the flow equations are switched off, energy equation is turned on, and step responses at all 16 output channels are recorded when 1 of the 6 heat sources are excited with a constant rate heat source, one-at-a-time. Overall, there will be 96 (6 × 16) step responses which will then be used to automatically extract an LTI ROM for this 2-cell unit model. The model was run GM's HPC using 64 cores. It took approximately 7 h to generate the data (step responses) to create the ROM. In order to verify the LTI ROM's accuracy, we compare the ROM generated step responses against full-field simulation data from CFD in Fig. 4.49a–f.

As seen in Fig. 4.49a, b, the match between the full-field CFD and LTI ROM is almost perfect. The same is true when the other four heat sources are excited one-

Fig. 4.48 A two-cell unit model with a cooling fin (blue) sandwiched between the two cells, tabs, and interconnects

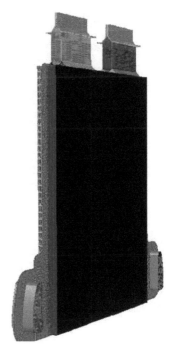

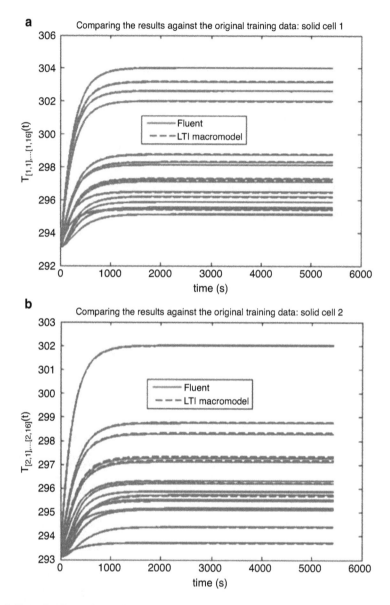

Fig. 4.49 (**a, b**) Comparison of the LTI ROM generated 16 step responses against the original CFD generated training data when Cell 1 (**a**) and Cell 2 (**b**) are excited – one-at-a-time. (**c, d**) Comparison of the LTI ROM generated 16 step responses against the original CFD generated training data when Tab 1 (**c**) and Tab 2 (**d**) are excited – one-at-a-time. (**e, f**) Comparison of the LTI ROM generated 16 step responses against the original CFD generated training data when Tab 3 (**e**) and Tab 4 (**f**) are excited – one-at-a-time

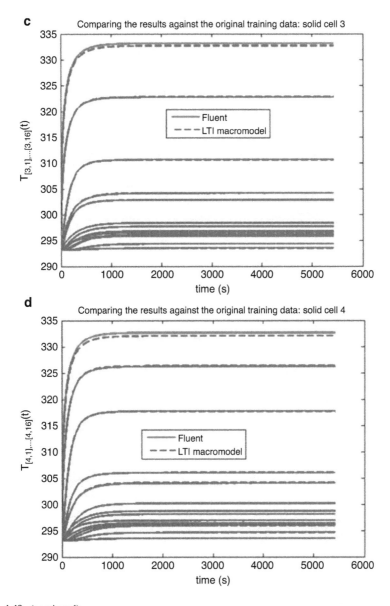

Fig. 4.49 (continued)

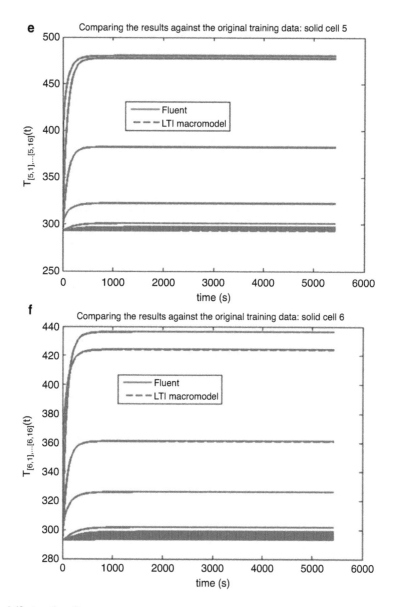

Fig. 4.49 (continued)

at-a-time and CFD generated step responses are compared with the LTI ROM-based model in the following figures.

Subsequent to verifying that LTI ROM reproduces the original training data from CFD satisfactorily, we use the same LTI ROM model to verify its accuracy against temperature profiles obtained using the full-order CFD/Thermal wherein the six input heat sources are applied as (1) constant 9.4 Watts on both cells and 1.7 Watts on all four tabs and (2) sinusoidal heat source on the cells given as $Q = A*\sin (2\pi*f*t)$ where A is the amplitude of 9.4 Watts and f is the frequency of 360 Hz. The comparison of temperature responses to these two heat source profiles obtained using LTI ROM and CFD/Thermal (Fluent) are shown in Figs. 4.50 and 4.51, respectively.

As observed in the preceding two figures (Figs. 4.50 and 4.51), the LTI ROM performs highly accurately when compared against full-order CFD/Thermal (Fluent) model at a fraction of the computational cost. In order to illustrate these savings, the full-order CFD/Thermal run in Fluent took approximately 1.5 h on an HPC (high-performance computer) using 64 cores, whereas it took only a couple of seconds for the LTI ROM model to reproduce those results – orders of magnitude faster without significant loss of accuracy.

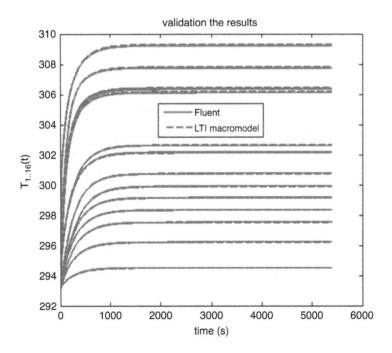

Fig. 4.50 Comparison of 16 temperature responses for constant heat source of 9.4 Watts on the two cells and 1.7 Watts on all four tabs, obtained using LTI ROM and CFD/Thermal (Fluent) models

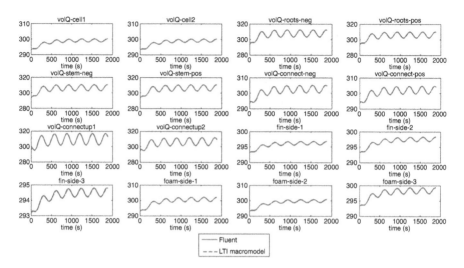

Fluent
--- LTI macromodel

Fig. 4.51 Comparison of 16 temperature responses to a sinusoidal heat source input to the two cells given by Q = 9.4*sin (2π*360*t) Watts and a constant 1.7 Watts on the four tabs obtained using LTI ROM and CFD/Thermal (Fluent) models

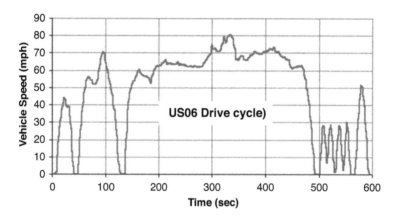

Fig. 4.52 US06 drive cycle

LTI Model Validation Results for US06

The earlier section on test setup describes in detail the layout of the module with cells, tabs, interconnects, bus-bars, cooling fins, ducts providing liquid cooling medium to the cooling fins, and thermocouple locations on cell surfaces. In order to validate the LTI ROM's performance against test data, we use the same LTI ROM that was verified against CFD/Thermal Fluent model in the earlier section. The module in the test was subjected to five back-to-back US06 drive cycle profiles. One of these profiles is shown in Fig. 4.52.

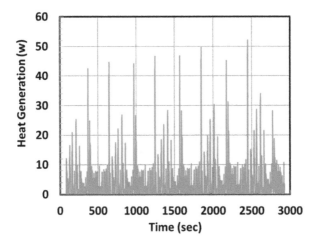

Fig. 4.53 Rate of heat generated in a cell of the module (2P-12S) undergoing five back-to-back US06 drive cycles using an ECM (equivalent circuit model)

Figure 4.52 shows how the vehicle speed varies with time for this US06 drive cycle. The sample period is almost 10 min at an average speed of 48.4 mph. Depending on a vehicle, its mass, rolling resistance, and aerodynamic characteristics, the power required can be easily estimated. This required power is converted into current for a given battery pack and is then used as an input to the ECM of the pack. This ECM provides the time-dependent heat generation in each cell of the module/pack for any given load that is needed as an input to both CFD/Thermal full-order model and LTI ROM. Figure 4.53 shows the heat generated in a cell for our module with 24 cells subjected to a load equivalent to 5 back-to-back US06 drive schedule for a compact car like Chevy Volt.

The ECM was generated using the standard HPPC (hybrid pulse power characterization) profile on the battery cell used in the module. The ECM's electrical performance is validated by comparing its terminal voltage against that recorded in the test for five back-to-back US06 drive profiles, as shown in Fig. 4.54.

As seen in Fig. 4.54, the ECM-predicted terminal voltage is quite satisfactory. Mismatch in ECM and test terminal voltage has a direct bearing on the accuracy with which the heat generation in a cell is predicted. Since there is no direct way of measuring heat generation in a cell or in a module, the next best way is to compare the ECM predicted heat generation against the difference in the rate of enthalpy change of the coolant in and out of the module. Since we know the rate of flow rate and the temperatures of the coolant at the inlet and at the outlet to the module, we can compute accurately the heat dissipated but the module as a whole to the coolant. This is shown in Fig. 4.55. Although we cannot directly compare the two heat rates, because of the large thermal time constant associated with the system, we can, however, compare the time-averaged values for the duration of the test. The ECM yields a time-averaged value of 2.88 Watts (Fig. 4.53), and the test gives 2.86

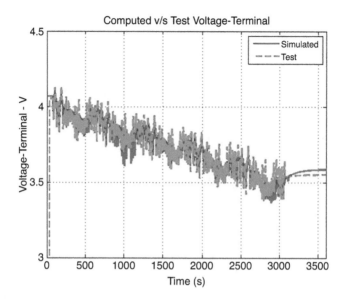

Fig. 4.54 Comparison of ECM computed terminal voltage of a cell in the module against that recorded in the test for five back-to-back US06 drive cycles

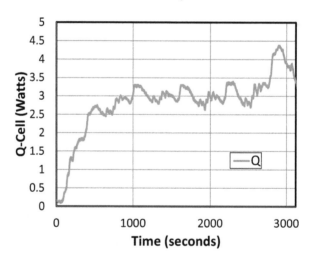

Fig. 4.55 Estimate of rate of heat generation in a cell indirectly using the difference in the rate of change in the enthalpy at the inlet and outlet of the module from the test data

Watts (Fig. 4.55). Hence, the ECM does a decent job of predicting heat generation in a cell (module/pack) – at least in a time-averaged sense.

Applying the ECM supplied heat generation in a cell for five back-to-back US06 (see Fig. 4.53) to the LTI ROM model, we can predict the temperature profiles at the six pre-defined locations on a two-cell unit model where the thermocouples were located. These LTI ROM generated temperatures are compared against the test data in these locations in Fig. 4.56a–f.

As Fig. 4.56a–f shows, the LTI ROM for thermal does a decent job of predicting the temperature profiles at all the six thermocouple locations on a cell in the module. Again, as mentioned earlier, these LTI ROM generated temperatures profiles for five back-to-back US06 profiles (50 min of real-time driving) were obtained in a few seconds.

In conclusion, indirect or black-box ROMs are highly effective and reasonably accurate when compared with test data. They are very accurate when compared with full-order CFD/Thermal Fluent generated data, as was shown in the ROM verification section.

4.5 Future Plans

As ANSYS leads the transfer of the newly developed technology to the industry, several logical continuation R&D activities can also be recommended to further increase the software's capability and appeal. One clear example is to evolve the cell-aging models into general-purpose, integrated features that would connect the deterministic, single-drive cycle simulations currently emphasized in ABDT to the stochastic service-life prediction that is ultimately needed in design. Another would be to integrate a more sophisticated model to address particle morphology, size distribution, surface modification, contact resistances, and mixture composition of active particles. The structural analysis capabilities in ANSYS' existing family of commercial products, although automatically tied loosely to the new battery tools through Workbench, were not utilized significantly in this project; a follow-on project could leverage that immense investment by extending the ABDT concept to microstructural models of resolved electrodes and/or macroscopic mechanical battery abuse scenarios.

The MSMD is recognized as an effective model framework for modular architecture linking interdisciplinary battery physics across varied length and timescales. By implementing MSMD in Fluent, the team has overcome challenges in modeling the highly nonlinear multi-scale response of battery systems. However, the inevitable nested iteration, ensuring self-consistency at each hierarchical level in the original MSMD, becomes a factor limiting computation speed.

In a separate concurrent project supported by DOE, NREL developed a new quasi-explicit nonlinear multi-scale multi-physics framework, the GH-MSMD. The new framework uses timescale separation and variable decomposition to eliminate several layers of nested iteration and keeps the modular MSMD architecture that is critical to battery behavior simulations. Fast electronic charge balance is differentiated from the processes related to slow ionic movements. During preliminary benchmark tests carried out at the electrode domain model (EDM) level, the GH-MSMD implementation demonstrated significant computational speed improvement compared to the original MSMD. One promising candidate to build

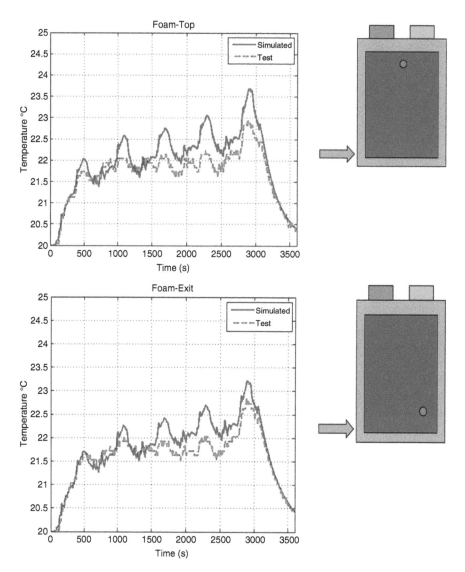

Fig. 4.56 (**a**) Comparison of LTI ROM generated temperature profile at the top (red dot) of the foam side on the cell surface against test data. (**b**) Comparison of LTI ROM generated temperature profile at the exit (red dot) of the foam side on the cell surface against test data. (**c**) Comparison of LTI ROM generated temperature profile at the inlet (red dot) of the foam side on the cell surface against test data. (**d**) Comparison of LTI ROM generated temperature profile at the top (red dot) of the fin side on the cell surface against test data. (**e**) Comparison of LTI ROM generated temperature profile at the exit (red dot) of the fin side on the cell surface against test data. (**f**) Comparison of LTI ROM generated temperature profile at the inlet (red dot) of the fin side on the cell surface against test data

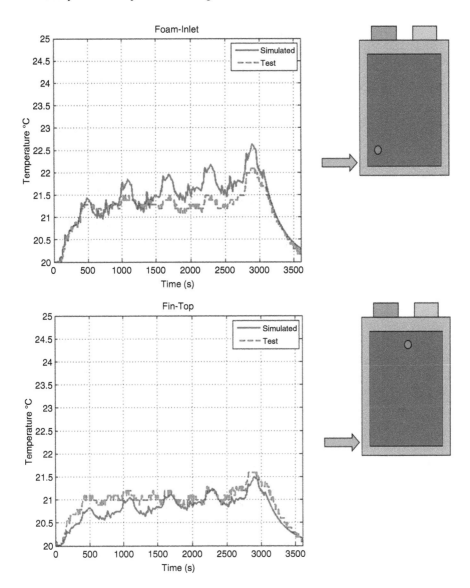

Fig. 4.56 (continued)

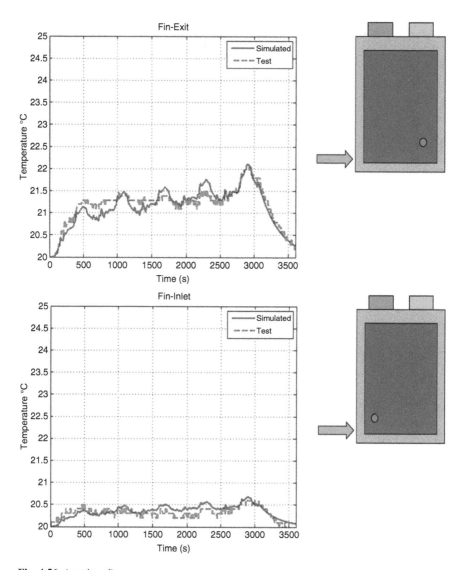

Fig. 4.56 (continued)

on the accomplishments of this project is therefore to implement GH-MSMD into the commercially deployed ABDT tool, potentially increasing computational speed of the pack-level simulation by a factor of 100.

Acknowledgments Original work performed under Subcontract No. ZCI-1-40497-01 under DE-AC36-08GO28308. This work would not have been possible without the support and guidance of many people. The authors wish to thank Brian Cunningham from the Vehicle Technologies Program of the DOE for funding support and guidance. We also wish to thank Gi-Heon Kim and Ahmad Pesaran from NREL for their contributions to the CAEBAT program by providing consultations as a technical monitor of the project. Finally, we express appreciation for the contributions from the following colleagues: Saeed Asgari, Steve Bryan, Jing Cao, Kuo-Huey Chen, Lewis Collins, Erik Ferguson, Amit Hochman, Sameer Kher, Genong Li, Shaoping Li, Justin McDade, Sorin Munteanu, Ramesh Rebba, Victor Sun, Dimitrios Tselepidakis, Michael Tsuk, Jasmine Wang, and Erik Yen.

Publications and Presentations

1. *Taeyoung Han, Gi-Heon Kim, Lewis Collins, "Multiphysics simulation tools power the modeling of thermal management in advanced lithium-ion battery systems," ANSYS Quarterly magazine "Advantage", 2012.*
2. *Taeyoung Han, Gi-Heon Kim, Lewis Collins, "Development of Computer-Aided Design Tools for Automotive Batteries-CAEBAT," Automotive Simulation World Congress (ASWC), Detroit, October 2012.*
3. *Xiao Hu, Scott Stanton, Long Cai, Ralph E. White, "A linear time-invariant model for solid-phase diffusion in physics-based lithium ion cell models.," Journal of Power Sources 214 (2012) 40–50.*
4. *Xiao Hu, Scott Stanton, Long Cai, Ralph E. White, "Model order reduction for solid-phase diffusion in physics-based lithium ion cell models," Journal of Power Sources 218 (2012) 212–220.*
5. *Meng Guo, Ralph E. White, "A distributed thermal model for a Li-ion electrode plate pair," Journal of Power Sources 221 (2013) 334–344.*
6. *Ralph E White, Meng Guo, Gi-Heon Kim, "A three-dimensional multi-physics model for a Li-ion battery", Journal of Power Sources, 2013.*
7. *Saeed Asgari, Xiao Hu, Michael Tsuk, Shailendra Kaushik, "Application of POD plus LTI ROM to Battery Thermal Modeling: SISO Case, to be presented in 2014 SAE World Congress.*
8. *Xiao Hu, Scott Stanton, Long Cai, Ralph E. White, "A linear time-invariant model for solid-phase diffusion in physics-based lithium-ion cell models.," Journal of Power Sources 214 (2012) 40–50.*
9. *M. Guo and R. E. White, "Mathematical Model for a Spirally-Wound Lithium-Ion Cell," Journal of Power Sources 250 (2014), also presented at the ECS meeting, spring 2014, Orlando, FL.*
10. *R. Rebba, J. McDade, S. Kaushik, J. Wang, T. Han, "Verification and Validation of Semi-Empirical Thermal Models for Lithium Ion Batteries," 2014 SAE World Congress, Detroit, MI.*
11. *G. Li, S. Li, "Physics-Based CFD Simulation of Lithium-Ion Battery under a Real Driving Cycle", Presentation at 2014 ECS and SMEQ Joint International Meeting, Oct 5–9, 2014, Cancun, Mexico.*
12. *G. Li, S. Li and J. Cao, "Application of the MSMD Framework in the Simulation of Battery Packs", Paper IMECE2014–39882, Proceedings of ASME 2014 International Mechanical Engineering Congress & Exposition, IMECM 2014, Nov 14–20, 2014, Montreal, Canada.*
13. *Y. Dai, L. Cai, and R. E. White, "Simulation and Analysis of Inhomogeneous Degradation in Large Format LiMn2O4/Carbon Cells," Journal of The Electrochemical Society, 161 (8), 2014.*
14. *T. Han, G. Kim, R. White, D. Tselepidakis, "Development of Computer Aided Design Tools for Automotive Batteries," ANSYS Convergence conference, Detroit, MI, June 5, 2014.*
15. *T. Han, M. Fortier, L. Collins, "Accelerating Electric-Vehicle Battery Development with Advanced Simulation," Aug 21, 2014, Webcast seminar organized by SAE International, http://www.sae.org/magazines/webcasts.*

References

1. Kim GH et al (2011) Multi-domain Modeling of lithium-ion batteries encompassing multi-physics in varied length scales. J of Electrochem Soc 158(8):A955–A969
2. ANSYS (2014) Fluent 15.0 user's guide. Ansys Inc.
3. Kwon K, Shin C, Kang T, Kim CS (2006) A two-dimensional modeling of a lithium-polymer battery. J Power Sources 163:151–157
4. Kim US, Yi J, Shin CB, Han T, Park S (2011) Modeling the dependence of the discharge behavior of a lithium-ion battery on the environmental temperature. J Electrochem Soc 158(5):A611–A618
5. Kim US, Shin CB, Kim C (2008) Effect of electrode configuration on the thermal behavior of a lithium-polymer battery. J Power Sources 180(2):909–916
6. Chen M, Rincon-Mora GA (2006) Accurate electrical battery model capable of predicting runtime and I–V performance. IEEE Trans Energy Convers 21(2):504–511
7. Doyle M, Fulle T, Newman J (1993) Modeling of Galvanostatic charge and discharge of the lithium/polymer/insertion cell. J Electrochem Soc 140(6):1526–1533
8. Cai L, White RE (2009) Reduction of model order based on proper orthogonal decomposition for Lithium-Ion battery simulations. J Electrochem Soc 156(3):A154–A161
9. Smith K, Wang CY (2006) Solid-state diffusion limitations on pulse operation of a Lithium ion cell for hybrid electric vehicles. J Power Sources 161:628–639
10. Li G, Li S (2014) Physics-based CFD simulation of Lithium-ion battery under the FUNS driving cycle. In: ECS transactions, October 5–9, Cancun, Mexico
11. Guo M, White RE (2013) A distributed thermal model for a Li-ion electrode plate pair. J Power Resour 221:334–344
12. Thomas K et al (2001) Measurement of the entropy of reaction as a function of state of charge in doped and undoped lithium manganese oxide. ECS J 148(6):A570–A575
13. Holman J (1976) Heat transfer. In: Section 4.2, "lumped heat capacity system", 4th edn. McGraw-Hill
14. ANSYS Battery Design Tool User's Manual, ANSYS, Inc., Version 1.0, 2014
15. Grimme EJ (1997) Krylov projection methods for model reduction, Doctoral dissertation, University of Illinois at Urbana-Champaign
16. Hatchard TD et al (2001) J Electrochem Soc 148:A755–A761
17. Kim GH, Pesaran A, Spotnitz R (2007) J Power Sources 170:476–489
18. Spotnitz RM et al (2007) J Power Sources 163:1080–1086
19. Smith K et al (2010) Int J Energy Res 34:204–215
20. Santhanagopalan S, Ramadass P, Zhang J (2009) J Power Sources 194:550–557
21. MacNeil DD, Dahn JR (2001) J Phys Chem A 105:4430–4439
22. Bayarri MJ et al (2007) Technometrics 49:138–154
23. Bayarri MJ et al (2007) Ann Stat 35:1874–1906
24. ASME PTC 60 (2006) ASME V&V 10 guide, New York
25. Han T, Tselepidakis D, White RE (2015) Development of computer aided design tools for automotive batteries. GM CAEBAT Project – Final report, prepared for subcontract No. ZCI-1-40497-01 under DE-AC36-08GO28308, October 2015

Chapter 5
Experimental Simulations of Field-Induced Mechanical Abuse Conditions

Loraine Torres-Castro, Sergiy Kalnaus, Hsin Wang, and Joshua Lamb

Abstract Mechanical failure of cell components is an important component to understanding safety outcomes for batteries subjected to external loading. Several key parameters governing the mechanical response at the component to the module level are discussed in this chapter. Specifically, nuances associated with experimental measurements of these properties – including development of test methods, fixturing the test articles, ways to minimize errors in measurements, and improving the repeatability of some of these experiments – are examined closely.

5.1 Introduction

Mechanical testing is well established within abuse testing to evaluate cells' susceptibility to deformation or intrusion. Most common are the use of nail penetration and crush tests that look primarily at the abuse response to mechanical damage [1, 2]. The original goal of these tests is predominantly to determine the battery failure response if an event does occur [3–6]. Later research made some efforts to look at mechanical damage as a stand in for internal short circuit testing, predominantly due to a lack of an accepted internal short circuit test [7–10]. A secondary concern in much of this testing has been a detailed look at the mechanical behavior of cells and batteries and the degree of mechanical damage needed to initiate failure.

The earliest mechanical tests developed were nail penetration tests, developed primarily due to the concerns of accidental puncture during manufacturing (e.g., accidental puncture with a nail gun) [2]. The lack of an accepted internal short circuit test however led to the technique to be widely used as a stand-in method

L. Torres-Castro · J. Lamb (✉)
Advanced Power Sources R&D, Sandia National Laboratories, Albuquerque, NM, USA
e-mail: jlamb@sandia.gov

S. Kalnaus · H. Wang
Oak Ridge National Laboratory, Oak Ridge, TN, USA

This is a U.S. government work and not under copyright protection in the U.S.;   187
foreign copyright protection may apply 2023
S. Santhanagopalan (ed.), *Computer Aided Engineering of Batteries*,
Modern Aspects of Electrochemistry 62,
https://doi.org/10.1007/978-3-031-17607-4_5

to determine internal short circuit susceptibility. Many researchers have detailed significant drawbacks to using nail penetration in this way [9–17]. The intrusion of a conductive nail causes widespread damage and a short circuit event through multiple cell layers [15] and is not particularly emblematic of mechanisms proposed for internal short circuit when compared to other methods [17]. The test method is still in wide use, however, partially due to the still extant concerns of mechanical intrusion during manufacturing or on-the-road incidents and also due to the relative ease of performing the test compared to other methods. Mechanical deformation has been used by researchers including Greve and Fehrenbach [18], Sahraei et al. [19], and researchers from Motorola [7, 8] finding that failures in these conditions typically arise from macroscopic damage to the electrodes, such as large cracks through the electrode jelly roll or delamination of electrode layers. While potentially closer to actual short circuits to nail penetration, they still are not quite able to produce true field failure conditions. Among other effects, this leads field failures to have a relatively high impedance (at least initially) and concentrate the related heat generation in a very small volume when compared to the failure caused by a sharp nail penetration [10–12, 14]. While the development of internal short circuit tests has been elusive, these methods taken together present tools to better understand the mechanical behaviors of lithium-ion cells and batteries.

The use of batteries in electric vehicles means that the risk posed by the battery during on-the-road incidents needs to be well understood. Many documented vehicle battery fires have involved some sort of crash or damage to the battery [20]. A robust mechanical understanding is commonly used to ensure that vehicles are reasonably safe during a crash scenario. The development of these models requires a general understanding of the mechanical properties and behaviors of the batteries themselves, which can be thought of as a composite material that is not well described by existing materials properties. The work presented here details mechanical tests used to better understand the mechanical behavior of lithium-ion cells and the limits that will initiate thermal runaway. This also provides data for the development and validation of mechanical behavior models presented in other chapters.

5.2   Experimental Design

All testing was performed on 5 Ah $LiCoO_2$-graphite commercial off-the-shelf (COTS) cells. Testing performed at 100% State of Charge (SOC) were charged at a 1C constant current rate followed by a current taper to 0.25 A at full voltage prior to mechanical testing. The majority of testing was performed at ambient temperature unless otherwise noted.

Mechanical testing of cells was performed using cylindrical impactors on single cells. Representations of the test orientation are shown in Figs. 5.1, 5.2 and 5.3. Testing was performed initially at 100% SOC to determine the deformation

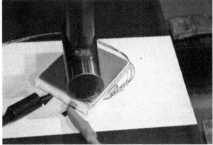

Fig. 5.1 Single-cell crush orientations showing crush parallel to cell tabs (left) and diagonal to cell tabs (right)

Fig. 5.2 Three-point bend test performed on single cells

Fig. 5.3 12-cell crush with testing fixtures built to constrain packs outside the direction of crush impact. Cylindrical crush into cell tabs shown (left) and fixture for crush into cell face/electrode layers shown (right)

necessary to achieve a thermal runaway failure. Subsequent testing and analysis was performed on cells at 0% SOC to the deformation where failure was observed.

Testing was performed at deformation rates of 0.01–10 mm/s with a 24.4 mm radius cylindrical impactor. Data collection rates were dependent upon the deformation rate used with a data collection rate of 1 Hz at 0.01 mm/s deformation up to a recording rate of 100 Hz at 10 mm/s deformation rate. The data collection rate was chosen to target a data rate of at least 10 points/mm of deformation during the active crush. Testing was performed at room temperature as well as at elevated temperatures of up to 120 °C. Mechanical testing was not performed above this temperature due to breakdown of electrolyte and other cell materials convoluting mechanical testing results.

Testing of 12-cell packs was performed in constrained fixtures as shown in Fig. 5.3. The constraint fixtures were used to minimize lateral deformation during the crush test of the cell. A semicylindrical impactor was used during crush into the cell tabs as shown, while a flat impactor was used to crush into the cell faces in Fig. 5.3 (right). All pack crush was performed at a speed of 2 mm/s with data recording at 10 Hz.

Pack testing was performed first on packs at 100% SOC to evaluate the level of deformation needed to initiate thermal runaway of the packs. Once this was determined, cells were evaluated at 0% SOC to provide post-test examinations of the mechanical deformation. This was evaluated after crush using both visual inspection and computed tomography (CT) imaging. All testing of 12-cell packs was performed at ambient temperatures.

5.3   Mechanical Behavior of Single Cells Approaching Thermal Runaway

Figure 5.4 shows the impact of temperature on fully charged 5 Ah pouch cells. While the typical operating ranges of lithium-ion batteries show little change in mechanical behavior, this initial work instead evaluated cells above normal operating temperatures. The cells were compressed at a rate of 0.1 mm/s with a 24.4 mm radius cylindrical impactor. Cells were tested at both 100% SOC and 0% SOC. 100% SOC cells were compressed until thermal runaway was observed, while cells at 0% SOC were compressed to the average deformation required to create thermal runaway. The current results show little change in the overall mechanical behavior of the cells up to 100 °C; however at 120 °C, we observed a significant reduction in the force required to deform the battery. We also observed a slightly higher displacement before thermal runaway occurred at elevated temperatures.

The observations here suggest that temperature would have minimal impact on the resilience of cells to mechanical deformation at more moderate temperatures. This shows little apparent change in the behaviors at lower temperatures suggesting that until 120 °C, we would not expect the susceptibility to mechanical crush to

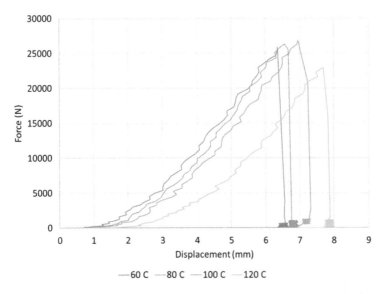

Fig. 5.4 Force vs displacement of fully charged 5 AH pouch cells. Little change is observed up to 60 °C; however a dramatic softening of the cell is observed at 120 °C. Crush in all cases was ended when thermal runaway events were observed in the cells

change significantly. On-going testing is looking at this elevated temperature testing at different rates and orientations to provide a complete picture on this susceptibility.

The next set of tests examined if alternate orientations and deformation rates of compressive force would have an impact in the mechanical behavior with results presented in Fig. 5.5. This shows compression along an axis parallel to the tabs (Fig. 5.5, left) and diagonally through the cell (Fig. 5.5, right). Both orientations show little effective change at the lowest rates, with little change from 0.01 to 0.1 mm/s in either orientation. The compression parallel to the tabs saw a more significant

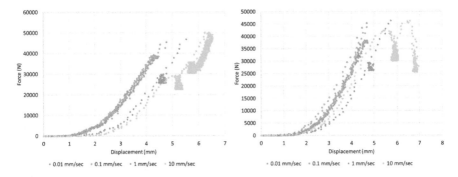

Fig. 5.5 Impact of strain rate at cell orientations parallel to cell tabs (left) and diagonal to cell tabs (right)

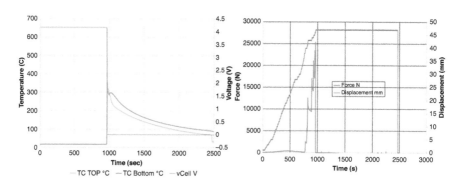

Fig. 5.6 Three-point bend testing of fully charged cell. No runaway was observed due to the bending of the cell. Runaway occurred when enough displacement occurred to begin compressing the cell with significant force

shift in the force at higher rates, with the force dropping at equivalent displacements when rates of 1 and 10 mm/s were applied. Changes were observed in the diagonal orientation as well; however these changes were less systematic than other observed results. The diagonal direction also saw some evidence of fracturing failure within the cell at higher levels of force, with sudden drops in the force as the electrode stack likely fractured.

Much of the mechanical deformation events observed in practice have been mechanical buckling events. Testing in the three-point bend fixture was used to evaluate this under both 100% and 0% SOC. This data is shown in Figs. 5.6 and 5.7. The first tests performed were on fully charged cells to determine at what point catastrophic failure is likely to occur when subjecting these cells to extreme mechanical deformation. Selected results from testing on fully charged cells are shown in Fig. 5.6.

In Fig. 5.6, deformation was applied in 1 mm increments with voltage and temperature monitoring to determine at what level of displacement thermal runaway occurred. Figure 5.6 shows the voltage, temperature, force, and displacement results versus time, while Fig. 5.7 (right) shows the force vs. displacement curve for the test. No runaway was observed during the initial bending. The observed thermal runaway occurred when enough deformation occurred to begin compressing the cell against the base of the fixture, with cell failure occurring due to compression of electrode layers rather than breakage from the bending force. Ultimately in this evaluation, compression of the cells was always required to initiate a thermal runaway event; bending of the cell itself did not cause sufficient damage to initiate cell failure.

The next step of this work was to determine the mechanical behavior of these cells under these conditions. The bend test was applied at strain rates from 0.01 mm/s to 10 mm/s to observe how the deformation was impacted under increasing rates of deformation. This data is shown in Fig. 5.8. This shows that there is little apparent strain rate impact at low levels of deformation; however some changes were seen in the yield points of the cells where a maximum force

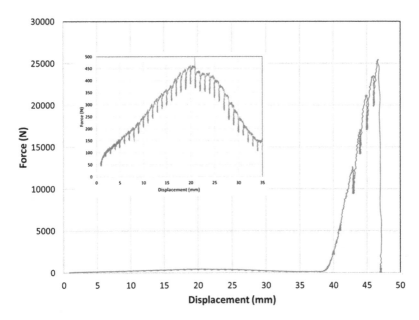

Fig. 5.7 Force vs displacement curve for test performed in Fig. 5.6

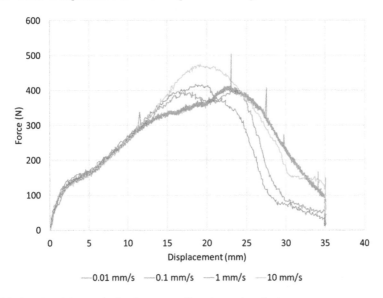

Fig. 5.8 Results of three-point bend tests on cells at increasing displacement rates

was reached and the cells exhibited irreversible plastic deformation. The data shows yield stresses that shift toward lower deformation levels and a higher overall force observed before the yield occurred as the deformation rate was increased.

The single cell work under multiple modes of deformation and failure shows little elasticity to the deformation rate at the rates that were tested here. Ultimately as well, it was observed that bending/buckling of the cell and its constituent layers were insufficient to cause thermal runaway on their own. Ultimately compression of the electrode was required to initiate a thermal runaway event.

5.4 Cylindrical Deformation into 12-Cell Module Stacks

Crush testing on representative 12-cell packs was performed to provide an experimental basis for how electric vehicle batteries may perform during a collision. This is coupled with CT imaging to better understand how the packs are deforming internally after undergoing a large-scale deformation. Selected results of this work are shown in Figs. 5.9, 5.10, 5.11 and 5.12.

12-cell packs are constructed by stacking commercial off-the-shelf 5 Ah pouch cells together in a close-packed configuration. Mechanical testing of these packs included both a strict compaction of the pack through the longest pack dimension perpendicular to the electrode stacks within the cell and a cylindrical impactor affecting four cells within the center of the pack with force applied parallel to the electrode stacks within the cell. These configurations cover cell failure through compaction of the electrode layers in the first case and buckling of the electrode

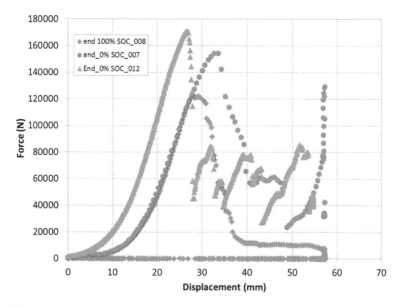

Fig. 5.9 Cylindrical crush into tabs of 12-cell packs comparing 100% and 0% SOC. 0% SOC cells were retained for CT evaluation to determine deformation behaviors at thermal runaway deformations

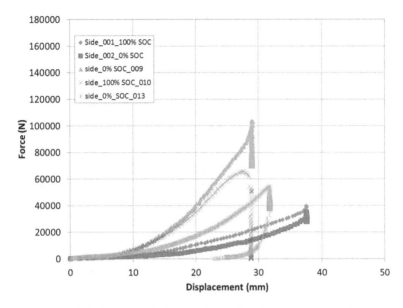

Fig. 5.10 Force and displacement into the face of 12-cell packs at 100% SOC and 0% SOC

Fig. 5.11 0% SOC cylindrical crush into electrode tabs. Insets (right) show the interior cutaways found using CT imaging

layers in the second. This is performed on both fully charged cells to determine the pack level failure response and discharged cells to better understand the mechanical deformation process. The results of this test on fully charged cell strings are shown

Fig. 5.12 0% SOC crush into cell face showing failure behavior. These cells show a shear fracturing through the cell body at failure cutting through multiple cells at the point of failure. This required significantly higher force to achieve

in Figs. 5.9 and 5.10, with Fig. 5.9 showing the results of compaction through the pack and Fig. 5.10 the results of the cylindrical buckling test.

This testing shows significantly different forces required to cause pack level failure. The different mechanical behaviors are evaluated on discharged cells with the aid of CT scanning of the cells. Figure 5.11 shows how buckling may progress through the cells as a result of the cylindrical impactor. This also shows areas of significant damage to the electrode layers as a result of this buckling, with electrodes pulling apart in some cases and becoming compacted together in others. This shows that in this mode, there were significant points where the electrode layers were compressed together, reinforcing the results observed above that compression within the cell was needed to cause a thermal runaway failure.

Figure 5.12 shows the visual results of the compression in the direction normal to electrode layers with representative of the data collected in Fig. 5.9. This shows that the sudden drops in applied force in that data are likely attributed to shear fracture events with sharp fractures observed cutting through multiple cells through the pack. In a fully constrained case like this, and with somewhat rigid constituent materials, there is little room for deformation of the electrode layers as the cell is compressed. This ultimately leads to fracturing of the cell layers that appears to match the deformation point of thermal runaway events at 100% SOC.

5.5 Summary

This work was performed as part of the USDOE CAEBAT program to understand how lithium-ion cells behaved under mechanical deformation and provide input to collision models being developed by partners within the program. Most directly, this work showed that the mechanical behavior of the completed cell materials does not significantly change under typical use conditions of lithium-ion batteries with little changes observed due to operating temperatures well above normal or from state of charge. While temperatures above 120 °C were not investigated, they are likely of little interest in this case as higher temperatures will run a high risk of thermal runaway due from the elevated temperature itself.

In the mechanical testing of cylindrical cells, in all cases no failure was observed until significant compression was applied to the cells. This was most notable in the three-point bend tests where no thermal runaway was observed in the low force portion where only bending of the cells was occurring. Cell failure was primarily driven at high force portions where enough deformation was performed to allow for significant compression of the cell.

This failure mode was evident in full pack tests as significant deformation of the cylindrical impactor was necessary before thermal runaway was observed. The cylindrical deformation into the cell tabs did not show a thermal runaway event until more than 50% of deformation into the original cell. At this point there was likely significant compression of the electrode layer together driving the high energy failure. Compression into the face of the cells was entirely a compression event; however this showed that fully constrained, these stacks of cells are also prone to shear fracturing as the force on the pack increases.

Acknowledgments Sandia National Laboratories is a multimission laboratory operated by National Technology and Engineering Solutions of Sandia LLC, a wholly owned subsidiary of Honeywell International Inc., for the US Department of Energy's National Nuclear Security Administration. Sandia Labs has major research and development responsibilities in nuclear deterrence, global security, defense, energy technologies, and economic competitiveness, with main facilities in Albuquerque, New Mexico, and Livermore, California. This paper describes objective technical results and analysis. Any subjective views or opinions that might be expressed in the paper do not necessarily represent the views of the US Department of Energy or the US Government.

This work was funded as part of the CAEBAT consortium funded by the US Department of Energy Vehicle Technologies Program.

References

1. Orendorff C, Lamb J, Steele LAM (2017) SNL abuse testing manual. Sandia National Laboratories, Albuquerque
2. Doughty DH, Crafts CC (2006) Freedom CAR electrical energy storage system abuse test manual for electric and hybrid electric vehicle applications. Sandia National Laboratories, Albuquerque
3. Balakrishnan PG, Ramesh R, Kumar TP (2006) Safety mechanisms in lithium-ion batteries. J Power Sources 155(2):401–414
4. Wu K et al (2011) Safety performance of lithium-ion battery. Prog Chem 23(2–3):401–409
5. Xia L et al (2011) Safety enhancing methods for Li-Ion batteries. Prog Chem 23(2–3):328–335
6. Scrosati B, Garche J (2010) Lithium batteries: status, prospects and future. J Power Sources 195(9):2419–2430
7. Cai W et al (2011) Experimental simulation of internal short circuit in Li-ion and Li-ion-polymer cells. J Power Sources 196(18):7779–7783
8. Maleki H, Howard JN (2009) Internal short circuit in Li-ion cells. J Power Sources 191(2):568–574
9. Orendorff CJ, Roth EP, Nagasubramanian G (2011) Experimental triggers for internal short circuits in lithium-ion cells. J Power Sources 196(15):6554–6558
10. Hyung YE et al (2011) Factors that influence thermal runaway during a nail penetration test Abstract 413. In: 220th ECS Meeting, Boston, MA
11. Jones HP, Chapin JT, Tabaddor M (2010) Critical review of commercial secondary lithium-ion battery safety standards. In: Proceedings 4th IAASS conference making safety matter, p 5
12. Loud J, Nilsson S, Du YQ (2002) On the testing methods of simulating a cell internal short circuit for lithium ion batteries. In: Das RSL, Frank H (eds) Seventeenth annual battery conference on applications and advances, proceedings. IEEE, New York, pp 205–208
13. Santhanagopalan S, Ramadass P, Zhang J (2009) Analysis of internal short-circuit in a lithium ion cell. J Power Sources 194(1):550–557
14. Wei H-B et al (2009) The comparison of li-ion battery internal short circuit test methods. Battery Bimonthly 39(5):294–295295
15. Lamb J, Orendorff CJ (2014) Evaluation of mechanical abuse techniques in lithium ion batteries. J Power Sources 247:189–196
16. Barnett B et al (2016) Successful early detection of incipient internal short circuits in Li-Ion batteries and prevention of thermal runaway. Meeting Abstracts MA2016-03(2):257
17. Finegan DP et al (2019) Modelling and experiments to identify high-risk failure scenarios for testing the safety of lithium-ion cells. J Power Sources 417:29–41
18. Greve L, Fehrenbach C (2012) Mechanical testing and macro-mechanical finite element simulation of the deformation, fracture, and short circuit initiation of cylindrical Lithium ion battery cells. J Power Sources 214:377–385
19. Sahraei E, Campbell J, Wierzbicki T (2012) Modeling and short circuit detection of 18650 Li-ion cells under mechanical abuse conditions. J Power Sources 220:360–372
20. Sun P et al (2020) A review of battery fires in electric vehicles. Fire Technol 56(4):1361–1410

Chapter 6
Abuse Response of Batteries Subjected to Mechanical Impact

Jinyong Kim, Anudeep Mallarapu, and Shriram Santhanagopalan

Abstract Electrochemical and thermal models to simulate nominal performance and abuse response of lithium-ion cells and batteries have been reported widely in the literature. Studies on mechanical failure of cell components and how such events interact with the electrochemical and thermal response are relatively less common. This chapter outlines a framework developed under the Computer Aided Engineering for Batteries program to couple failure modes resulting from external mechanical loading to the onset and propagation of electrochemical and thermal events that follow. Starting with a scalable approach to implement failure criteria based on thermal, mechanical, and electrochemical thresholds, we highlight the practical importance of these models using case studies at the cell and module level. The chapter also highlights a few gaps in our understanding of the comprehensive response of batteries subjected to mechanical crash events, the stochastic nature of some of these failure events, and our approach to build safety maps that help improve robustness of battery design by capturing the sensitivity of some key design parameters to heat generation rates under different mitigation strategies.

6.1 Introduction

Mechanical integrity and threshold for impact tolerance of cell components are of significant interest to battery designers, with their increase in cell sizes and battery volumes in various applications, especially in electrical vehicles [1–3]. Sensitivity to mechanical load, which might originate due to various factors internal or external to the cell casing, has a likelihood of introducing a localized change in impedance and/or a short circuit. The breach of mechanical failure threshold often triggers a high-rate discharge. Rapid exothermal decomposition reactions, commonly referred

J. Kim · A. Mallarapu · S. Santhanagopalan (✉)
National Renewable Energy Laboratory, Golden, CO, USA
e-mail: Shriram.Santhanagopalan@gm.com

This is a U.S. government work and not under copyright protection in the U.S.;
foreign copyright protection may apply 2023
S. Santhanagopalan (ed.), *Computer Aided Engineering of Batteries*,
Modern Aspects of Electrochemistry 62,
https://doi.org/10.1007/978-3-031-17607-4_6

to as thermal runaway, followed by cell venting and rapid dissipation of energy ensue. Understanding what factors influence the outcome of mechanical loading, and how the cell components subsequently interact, is critical to develop mitigation strategies to prevent localized failure events from propagating across multiple cells. Several standards have been proposed over the years for various applications [4]. Apparently minor changes in mechanical failure thresholds result [5] in exponential differences on the electrical and thermal response: for instance, a 60 μm increase in displacement during a battery crush following an indentation can result in an order of magnitude increase in the short-circuit current. These results predict a drastic reduction in response time to contain failure following a runaway event.

Progression of structural failure across the jellyroll is a crucial first step that determines the mode of internal failure and how the electrodes respond electrically and thermally following a short circuit. Experimental results [6] show a combination of multiple failure modes, including rupture of the metal foils that constitute the current collectors, melting of aluminum, multi-layer mechanical shear, secondary short circuit from fragments, or dislodged active material. The short-circuit resistance is in turn influenced by these factors. It is equally important to (a) understand the independent failure modes and (b) the electrical, thermal, electrochemical, and mechanical interactions across different cell components. For instance, Marcicki et al. [7] show that the contact duration between the mechanical indenter and the cell is a crucial factor in predicting the voltage drop during a high-speed impact. They further show results that indicate a higher propensity for thermal propagation, following a larger voltage drop.

Even before considering the safety response, several factors complicate modeling mechanical response of composite battery electrodes. The response of electrodes (and separators) during compression was found to be [8] different from that under tension. There is a gradual increase in stiffness during the compression tests as the void volume within the electrodes get eliminated [10]. Mechanical response during the tensile results was dominated by the properties of the current collectors. Thus, whereas the compressive response of cell components can be modeled as foam materials, the tensile response must be modeled like that of a thin metal foil. In-plane measurements of mechanical properties are in general not available due to the complications arising from sample buckling; but Lai [11, 12] has reported formation of shear bands in plate pairs from graphite/LiFePO$_4$-type cells and compared the response of cell components to in-plane versus out-of-plane compression. The tensile failure strain of the cathode and anode current collectors are lower than that of the respective metals (i.e., a sheet sample of aluminum or copper with the same thickness). This is attributed to the damage induced by intrusion of active particles into the metal foil [13]. There are differences in properties between dry electrodes and wet cell components [9, 14]. Wierzbicki and co-workers have extensively characterized mechanical behavior of pouch and cylindrical format cells under different loading conditions [10, 15, 16]. Some research groups have reported linear variation in the modulus or failure strain of electrodes with state-of-charge (SOC) or chemical composition [17, 18], although in our work, we have found the mechanical properties to be more sensitive to aging conditions [19] or temperature of operation

[20]. Similarly, strain-rate dependence of separators [9, 14, 21, 22] and electrodes [23] have been reported using extensometer-type quasi-steady-state measurements. As we show in a subsequent section, the response of cell components under high-speed impact (corresponding to a vehicle-crash scenario) varies considerably from the trend based on quasi-steady-state measurements reported in the literature.

In this chapter, we review different approaches to incorporate mechanical response of cell components into an electrochemical model, starting with calibration of some of the model parameters from experimental measurements. We highlight a few key challenges faced by the battery community in developing a comprehensive model that reliably captures failure outcomes under a variety of different loading scenarios. We discuss in detail, the numerical implementation of these models and associated modifications from the conventional battery model development protocols. We compare different failure modes against experimental results and their implications for propagation of failure from a point source across multiple cells. We finally discuss the limitations in experimental methods and propose alternate ways to develop criteria for failure analysis.

6.2 Mechanical Modeling Framework

Some of the early efforts on modeling mechanical failure. In cell components were spearheaded by Sahraei and Wierzbicki [2, 15, 24–26]. Similarly, Avdeev and Mehdi [27, 28] considered lumped macro-scale models to represent deformation of lithium-ion cells. As noted above, dynamic-response and strain-rate effects were incorporated by Xu et al. [17, 29] The macro-scale mechanical response of the battery is very useful in determining the load capacity of the structure, but is not sufficient to understand the different failure mechanisms or the interplay between mechanical failure and onset of electrical short circuit. Similarly, the occurrence of mechanical failure, in and of itself, does not result in electrical or thermal faults. To study these interactions in a methodic fashion, we need a comprehensive understanding on the constitutive and failure properties of the cell components (i.e., active material coating, separator, current collector). This framework should account for different trigger mechanisms such as voltage failure due to sudden drop in cell impedance, physical contact due to mechanical failure, or deformation due to change in mechanical properties with temperature. Thus far, the interaction between failure modes or the constitutive representation of the porous structure had not been characterized adequately; but providing a framework was imperative to model the progressive failure in the stack or jellyroll, under external loading.

We presented [30, 31] an alternate representation of the cell components in a mechanical model, tentatively based on the porous-electrode representation [32] where the individual cell components are retained for tracing their physical significance, albeit allowing for simplification of the geometry using "effective properties." The representative-sandwich model [31, 33–35] thus explicitly accounts for each cell component and predicts the electrical short circuit and temperature ramp due to a mechanical crush. Saharei et al. [36] modeled the failure sequence of various cell components under different loading scenarios using the representative-sandwich model. In both experimental and numerical studies, we found that the initiation of structural failure is mainly driven by the deformation of the electrode active material coatings due to their much lower tensile failure strains, followed by rupture of the current collector (especially on the anode) which in turn resulted in localized accumulation of stresses on one side of the separator layer [35, 37, 38].

6.2.1 Constitutive Properties for the Mechanical Model

As elaborated in our previous work [31], the prediction of short circuit induced by mechanical failure is contingent upon accurate representation of progressive failure across cell components. This task necessitates explicit modeling of the response of individual cell components. Mechanical modeling of crush events in batteries thus requires the development of representative constitutive models for the cell components. Prior reports [8] already provide some preliminary insights on the tensile and compression performance at the component level such as properties of electrodes and separators. The next step is to verify the suitability of these component models to predict multi-stage deformations in stacks of electrodes and separators alternating each other, as part of validating these models. This section outlines our approach to estimate mechanical parameters accounting for the porosity in the active material layer and to develop expressions that describe stress-strain response and failure properties for the porous coatings. We summarize the constitutive properties and propose suitable material models and their implementation in a finite element simulation.

We calibrated the models using test data from stacks of individual cell components under tension and compression; these models were then used to predict response in a stack configuration that mimics an actual cell, under indentation-induced failure conditions. Beyond the homogenized models previously available in the literature, we discuss modeling approaches for implementing the component-level models in battery crush simulations, explicitly simulating the deformation and damage behavior of electrodes, due to internal stresses that develop in the cell environment. These experimental and numerical results provide insights on improving the structural strength of lithium-ion battery and optimizing the design of safe battery structures as discussed in subsequent sections.

6.2.1.1 Sample Preparation and Experimental Setup for Mechanical Characterization

We harvested samples from a 15 Ah commercial graphite/NMC cell, with chemistry and thickness information summarized in Table 6.1. The cells were opened within

Table 6.1 Chemistry and thickness information for the cell components[a]

Component	Material	Thickness (μm)
Anode current collector + coating	Graphite	123
Anode current collector	Copper	14
Cathode current collector + active material	$LiNi_{0.33}Mn_{0.33}Co_{0.33}O_2$ (NMC)	92
Cathode current collector	Aluminum	20
Separator	PP/PE/PP	20

[a]Parameters values are from Ref. [35]

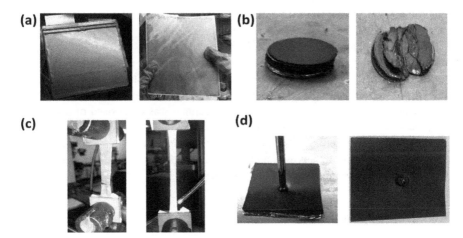

Fig. 6.1 (**a**) Destructive physical analysis (DPA) of a pouch format lithium-ion cell was conducted to harvest samples for evaluation of mechanical properties of the cell components. (**b**) Electrode samples used to measure properties under compression: multiple layers of each electrode were stacked to avoid effects of substrate interference. Results showed that the coatings on the electrodes failed before the current collectors. (**c**) Electrode samples used to measure mechanical response under tension: failure across mid-section of the samples was used to ensure that edge effects due to the fixture were minimal. Separator samples showed necking followed by plastic deformation. (**d**) Electrode samples used for indentation were sized such that the sample dimensions were at least six times larger than the indenter dimension to avoid edge effects. The indentation test results were used to validate component-level models populated with mechanical properties measured under compression and tension. All sample dimensions are listed under Sect. 6.2.1.1

an argon-filled glovebox, and proper care was taken to prevent short circuit or dry out of the electrodes, before and after test under different loading conditions.

Cell components harvested for these tests after destructive physical analysis are shown in Fig. 6.1. After cell tear down, each cell component is isolated and stored under a solvent (diethyl carbonate). The length and width of the cell were 190 mm and 210 mm, respectively. Right before the tests, the test samples were cleaned, taken out, and cut into shapes of required dimensions for mechanical tests. The representative images of the samples are shown in Fig. 6.1b–d.

A universal test machine with a 5kN load limit was used for quasi-static characterization. Force and displacement data were monitored during the tests to obtain the constitutive properties and failure strain of the samples. For tensile tests, the cell components were cut into $4'' \times 0.4''$ rectangular shapes using a specimen die. The gage section was set at $2''$ by tabbing $1''$ on each end using vice clamps. Multiple repeats of constant displacement rate tests between 0.2 and 2 mm/min were employed to subject the samples to uniaxial tension until fracture. The samples were also subject to tension under different orientations and to multi-axial tension tests in separate experiments (not shown here). The electrode properties were direction independent, whereas the separator membranes showed higher tensile strength along the machine direction compared to the transverse direction. These are essential qualities of the membrane to maintaining adequate balance between mechanical strength and resistance to heat-induced shrinkage.

The compression test specimen were circular pieces with $0.25''$ diameter, punched out using a tube-punch. To avoid interference from the baseplate, multiple layers of the same cell component (e.g., coated cathode foils) were stacked in the thickness direction during each experiment and installed between two steel plates connected to the test fixtures. Each anode specimen was comprised of a stack of 20 layers, while the separator specimen contains a total of 96 layers. To eliminate inconsistencies at lower initial loads, due to air gaps between the layers within each specimen, a preload of 10 N was applied. The compression tests were terminated once the sample fractured or upon reaching the load limit.

6.2.1.2 Tensile Response

The tensile response for each cell component (Fig. 6.2a, c and e) for the anode, cathode, and separator, respectively) shows good repeatability in terms of both the slope of the curves and the ultimate failure strain. The test results represent the elastic-plastic tensile behavior of cell components and capture the brittle fracture of electrodes and compliance of polymer separators. Following earlier reports by Sahraei et al. [8] and Lai et al. [11, 12], we use an isotropic linear-hardening elastic-plastic model for the electrodes:

$$\sigma = \begin{cases} E\varepsilon & if\ \sigma < \sigma^y \\ \sigma^y + E_t \left(\varepsilon - \frac{\sigma^y}{E}\right) & if\ \sigma \geq \sigma^y \end{cases} \tag{6.1}$$

where σ, ε, and E are the tensile stress, tensile strain, and tensile modulus, respectively. E_t and σ^y are the tangent modulus (defined as the stiffness of the material after transitioning to plastic deformation) and yield stress (the critical stress value at which the initiation of plastic deformation happens). Figure 6.2 also shows the fits for the stress-strain curves of the anode and cathode coatings on the respective current collectors, using Eq. (6.1). Subsequent empirical modifications to the yield-stress criterion were proposed by Fink et al. [19], to improve quality of fits, especially in aged electrode samples.

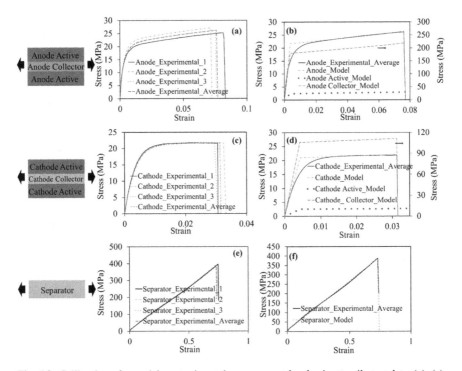

Fig. 6.2 Calibration of material properties at the component level using tensile test data: (**a**), (**c**), and (**e**) show repeatability of the experimental data in the composite anode, composite cathode, and separator, respectively; (**b**), (**d**), and (**f**) show fits to model equations and extraction of active material properties which cannot be directly measured in experiments. (Figure modified with permission from Ref. [35])

Battery electrodes are layered thin-film composites of active material coatings on metal foils. The coatings are in turn a porous composite of active particles and polymer binders whose contributions to the tensile strength of the electrode are minimal. Structural discontinuity within the porous layer results in the mechanical loads being transferred mainly through the current collector under uniaxial tension; however, deformation of the coatings due any shear stress transferred across the interface (i.e., peeling or disintegration of the active material coatings) is mainly attributed to differences in tensile strengths between the coatings and the foils. Experimentally measuring properties of the coatings are challenging, but it is important to delineate the strength of the coatings from that of the foils to understand safety response of the cell. To estimate the contributions of the coatings versus the current collectors, we employed a linear relationship to relate component contributions to the effective response of the electrode:

$$F_e = \alpha F_a + (1 - \alpha) F_c \qquad (6.2)$$

where F_e, F_a, and F_c are the tensile force experienced by the electrode, the active material, and the current collector, respectively. The fractional contribution of the active material coating to the tensile resistance is denoted by the parameter α. To simulate more complex interactions using a locally linearized response, α can be specified as a function of strain. Knowing the cross-sectional areas (sample width x thickness), we generated the stress-strain curves for the coatings and the current collectors for each electrode (Fig. 6.2b, d).

Tensile failure for the electrodes coincides with that of the current collectors. The coating or the active material has higher toughness values (as deduced from the higher tensile failure strains in Fig. 6.2b, d) than the current collectors. These results are summarized in [33]. Similar findings were reported by Wang et al. [38] In their study, the current collectors were fractured even though the active material was largely intact. This is attributed to the porous structure of the coatings and good compliance properties of the polymers that constitute the binder. The foils are reported to be relatively brittle ($\varepsilon_f \leq 10\%$) [23, 39, 40]. The foils are further weakened during the calendaring process. Calendaring reduces the mechanical strength of current collectors significantly and results in earlier onset of failure in current collectors compared to sheet metals of comparable thickness.

6.2.1.3 Compression Response

Unlike metals, laminated porous composites exhibit different mechanical properties under compressive versus tensile loading conditions. The failure modes under different loadings are also different. We employed the same set of properties for the current collectors under both tension and compression and estimated properties of the porous layers. We also assumed that the compression response is isotropic since in-plane measurements under compression are not practical. Figure 6.3a, c and e shows the component-level compression responses. Both the electrodes exhibited a two-stage deformation, starting with the gradual level off in the effective modulus corresponding to the initial porosity compaction and a subsequent linear response [11, 12].

Measured compression data are shown in Fig. 6.3. We employed a two-stage model to simulate the compression response of the active layer and separators. This model assumes that the Young's modulus E of the stiffening stage varies exponentially with the strain ε from initial modulus E_0 to a fully compacted value E_{max}.

$$E = \begin{cases} E\beta(\varepsilon - \varepsilon_p)_{max} & \varepsilon < \varepsilon_p \\ E_{max} & \varepsilon \geq \varepsilon_p \end{cases} \tag{6.3}$$

$$E_0 = E_{max}e^{-\beta\varepsilon_p} \tag{6.4}$$

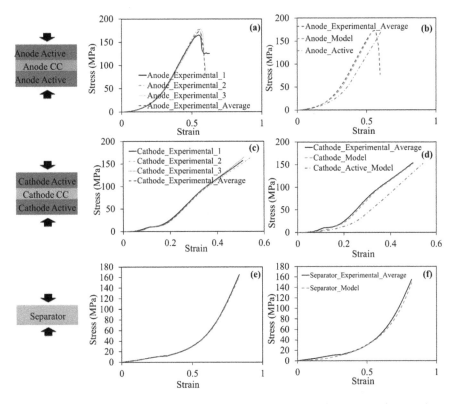

Fig. 6.3 Calibration of material properties at the component level using compression test data: (**a**), (**c**) and (**e**) show repeatability of the experimental data in the composite anode, composite cathode, and separator, respectively; (**b**), (**d**), and (**f**) show fits to model equations and extraction of active material properties which cannot be directly measured in experiments. (Figure modified with permission from Ref. [35])

In Eqs. (6.3) and (6.4), ε_p corresponds to the strain when the material is fully compacted. It can be approximated to the initial porosity of the respective layer. The fitting parameter β determines the increasing gradient of Young's modulus against strain in the stiffening stage (see Fig. 6.3a, c). A higher value of β indicates a slower increase of modulus with strain. By integrating Eq. 6.3, we calculate the corresponding stress-strain relationship:

$$\sigma = \int_0^\varepsilon E d\varepsilon = \begin{cases} \dfrac{E_{\max}\left(e^{\beta\varepsilon}-1\right)}{\beta e^{\beta\varepsilon_p}} & \varepsilon < \varepsilon_p \\ \dfrac{E_{\max}\left(1-e^{-\beta\varepsilon_p}\right)}{\beta} + E_{\max}\left(\varepsilon-\varepsilon_p\right) & \varepsilon \geq \varepsilon_p \end{cases} \qquad (6.5)$$

For the electrodes, which are comprised of active material coatings on either side of the current collector, we consider the effective through-thickness stress (σ_{eff}) to be the same for the three layers:

$$\sigma_{eff} = \sigma_{active\ mat} = \sigma_{collector} \tag{6.6}$$

To simulate the mechanical behavior of active material coatings, β and E_{max} are the only two unknown parameters. The effective strain (ε_{eff}) across the thickness is the sum of the volume (thickness) averaged strains in each component:

$$\varepsilon_{eff} = \nu_{active\ mat}\varepsilon_{active\ mat} + \nu_{collector}\varepsilon_{collector} \tag{6.7}$$

where $\nu_{actives}$ and $\nu_{collector}$ correspond to ratio fractions of the respective components. The procedure to deduce the parameters to simulate the compressive response of cell components, and the corresponding parameter values are detailed in [33].

6.2.1.4   Failure Response

Accurate material fracture criteria and damage evolution models for the components are necessary to predict mechanical failure in a battery cell. Failure strength under compression for the current collectors and the separator are much higher compared to that of active materials. A strain-based failure criterion is selected with the assumption that both anode and cathode active materials have similar damage behavior. Damage accumulates once the strain exceeds an initial threshold (ε_f^0) according to the following damage equation:

$$D = \left(\frac{\varepsilon}{\varepsilon_f}\right)^{\phi} \tag{6.8}$$

With accumulation of damage, material response is softened resulting in a reduction of stress. The stress status of compressive-damaged active material is represented as:

$$\sigma^* = \sigma \left(1 - \left(\frac{D - D_c}{1 - D_c}\right)^{\varphi}\right) \forall D > D_c \tag{6.9}$$

6.2.2   Numerical Implementation of the Mechanical Models

All mechanical models were implemented in LS-DYNA, since this was the tool of choice used by several original equipment manufacturers (OEMs) for crash-safety simulations at the vehicle level. Other variants of the tool, as well as other commercial explicit finite element simulation tools, are available. For the cell components, the MAT_HONEYCOMB material model was selected from the LS-DYNA library [15, 34, 41–44]. Anisotropic properties of the separators were included. The current collectors were modeled using elastic-plastic isotropy. Experimental datasets were imported into MAT_HONEYCOMB. Compressive failure strain ε_f from the tests was converted to the volumetric strain and applied via MAT_EROSION. When the strain reaches the failure threshold values, mesh elements are deleted. Owing to symmetry, the response from only a quarter of the experimental setup was simulated. Perfect bonding was implemented where no delamination was observed experimentally; thus active material and current collector interface was modeled using shared nodes. The indenter was a rigid sphere with an assigned velocity. Kinetic energy loss was minimal; so a time-scaling scheme was adopted to simulate various impact velocities.

6.2.3   Validation at the Components Level

Validation studies were conducted using a 4 mm diameter blunt rod to perform indentation tests on electrode layers. The indentation test specimen were 30 mm × 30 mm square pieces with five plate pairs comprised of separator/cathode composite/separator/anode composite layers. The punch was located over the center of the specimen. A preload of 2 N was used. Tests were conducted at a constant displacement rate of 0.2 mm/min. The end of test criterion was the fracture of the sample, recorded as a continuous drop in the measured force. Crack patterns were inspected using an optical microscope.

Experimental results in Fig. 6.4 show distinct peak forces representing individual cell component failure on the load-strain curve. Beyond the second peak, the

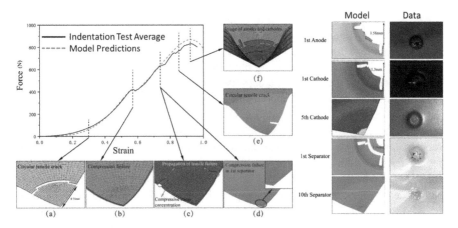

Fig. 6.4   Component-level validation of the mechanical models: using the compression and tensile data for the cell components, indentation using a blunt rod with 4 mm diameter was simulated and compared against experimental results. In addition to a reasonable comparison against the data, the model results were also able to capture failure modes within each cell component that trigger mechanical failure in each layer. (Figure reproduced with permission from Ref. [45])

specimen fracture is complete, and the force continuously reduces thereafter. The first peak load corresponds to compressive damage of the anodes, followed by a similar response of the cathode. Following this, radial tensile cracks evolve starting from directly below the indenter. These crack patterns match experimental observations. Thus, these simulations not only predict the onset of different failure mechanisms but also explain propagation of damage within the stack. A detailed discussion of damage evolution is provided by Zhang et al. [45]

6.2.4   Scaling Up to Cell- and Pack-Level Simulations

The next step in simulating the abuse response of batteries subject to mechanical loads is to scale up results from the individual layers or stacks of layers to the cell-level and higher length scales. Maintaining reasonable time steps while tracking deformation under high aspect ratios is critical to strike the right balance between keeping the computational burden reasonable and the simulations stable. As noted in the previous section, simulation times for the indentation response of five plate pairs over 65 hours call for simplification of the geometry without losing the fidelity of accounting for the various failure modes across each component. There are reports of efficient electrochemical-thermal simulations at the module and/or pack scale [46].

6.2.5   A Representative-Sandwich Model

There is a compromise between a fully homogenized jellyroll and the high-fidelity representation of individual electrodes or current collectors using mesh elements with aspect ratios over five orders of magnitude. We introduced a representative sandwich (RS) to define a repeating block as shown in Fig. 6.5. Each RS contains a cathode active material layer, a cathode current collector layer, an anode active material layer, an anode current collector layer, and one separator layer.

In this example, the whole pouch cell consisted of about 165 layers (or 40 representative sandwiches). One simplifying approximation is to represent the 40 representative sandwiches as one single equivalent RS that has proportionately higher failure strains, depending on the number of layers. This is accomplished for the case where each layer has a proportionately larger thickness (as shown on right side of Fig. 6.5). An alternate approach is to vary the number and thickness of RS-layers selectively to capture localized damage. Comparison of different simplifications and the impact of model order reduction on the mechanical accuracy as well as computational efficiency are discussed in detail, in [45].

One of the key challenges in employing a homogenized model is to maintain our ability to retain material properties for deformation and failure criteria for individual cell components. Next, we outline a procedure to calculate the effective

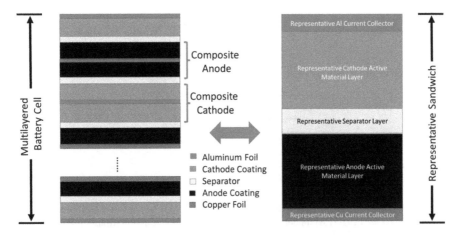

Fig. 6.5 Development of the representative-sandwich (RS) model geometry: to overcome difficulties with poor aspect ratios in mesh elements subject to mechanical deformation, and to improve numerical efficiency, the different layers within the cell are approximated by a single representative sandwich of equivalent thickness while retaining the individual failure criteria for the layers

properties such as components of the moduli in the stiffness matrix for the composite jellyroll, based on stress-strain measurements in the individual layers. For illustrative purposes, the analysis presented here is conducted using a single, solid, anisotropic, linear elastic, rectangular element under compression. A Cartesian coordinate system (x, y, z) is used, and the plate is in the positive z-region. The linear constitutive equations of a homogeneous elastic material can be written as:

$$\sigma_{ij} = C_{ijkl}\varepsilon_{kl} \tag{6.9}$$

where σ_{ij} and ε_{kl} are the stress and strain components and C_{ijkl} the elastic stiffness matrix. The stress/strain components can be partitioned into in-plane and out-plane vectors (corresponding to the subscripts $\parallel$ and $\perp$, respectively). For example, the stress components are organized as follows:

$$\boldsymbol{\sigma}_{\perp} = \{\sigma_{33}, \sigma_{23}, \sigma_{13}\}^{T} \quad \boldsymbol{\sigma}_{\parallel} = \{\sigma_{11}, \sigma_{22}, \sigma_{12}\}^{T} \tag{6.10}$$

In Eq. (6.10), the superscript T denotes the matrix transpose. Subscripts 1, 2, and 3 indicate the in-plane x and y and the through-plane z directions, respectively (Fig. 6.6). Then, the constitutive Eq. (6.9) can be equivalently written in the following concise matrix form:

$$\boldsymbol{\sigma}_{\parallel} = \boldsymbol{C}_{\parallel}\boldsymbol{\varepsilon}_{\parallel} + \boldsymbol{C}_{\times}^{T}\boldsymbol{\varepsilon}_{\perp} \tag{6.11}$$

$$\boldsymbol{\sigma}_{\perp} = \boldsymbol{C}_{\times}\boldsymbol{\varepsilon}_{\parallel} + \boldsymbol{C}_{\perp}\boldsymbol{\varepsilon}_{\perp} \tag{6.12}$$

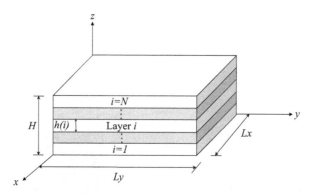

Fig. 6.6 Coordinate system used in the development of the effective constitutive properties for the representative-sandwich (RS) model. (Figure reproduced with permission from Ref. [34])

where $C_\perp$, $C_\parallel$, and $C_\times$ are the generalized Voigt matrices given by:

$$C_\parallel = \begin{bmatrix} C_{11} & C_{12} & C_{16} \\ C_{12} & C_{22} & C_{26} \\ C_{16} & C_{26} & C_{66} \end{bmatrix} \quad C_\perp = \begin{bmatrix} C_{33} & C_{34} & C_{35} \\ C_{34} & C_{44} & C_{45} \\ C_{35} & C_{45} & C_{55} \end{bmatrix} \quad C_\times = \begin{bmatrix} C_{13} & C_{23} & C_{36} \\ C_{14} & C_{24} & C_{46} \\ C_{15} & C_{25} & C_{56} \end{bmatrix}$$

$$(6.13)$$

Now we consider a jellyroll consisting of N layers with each layer comprised of homogeneous material. We assume that i) the displacements and tractions are continuous across each interface and ii) that the total thickness of the plate (H) is much lower than the in-plane dimensions (L_x, L_y). Then the in-plane strain in each layer can be approximated to the effective strain:

$$\bar{\varepsilon}_\parallel = \varepsilon_\parallel^{(1)} = \varepsilon_\parallel^{(2)} = \cdots = \varepsilon_\parallel^{(N)} \qquad (6.14)$$

The over bar represents effective properties of the layered composite and the superscript (i) the quantities associated with layer i ($i = 1, 2, \ldots, N$). Each of these layers can represent multiple computational elements within the same battery component or different physical battery components (anode, separator, cathode, etc.). The effective through-plane strain is calculated as the volume (thickness) average of the strain in the N layers:

$$\bar{\varepsilon}_\perp = v^{(1)} \varepsilon_\perp^{(1)} + v^{(2)} \varepsilon_\perp^{(2)} + \cdots + v^{(N)} \varepsilon_\perp^{(N)} \qquad (6.15)$$

where $v^{(i)} = \frac{h^{(i)}}{H}$ is the volume fraction of the i-th layer and $H = h^{(1)} + h^{(2)} + \ldots + h^{(N)}$ is the total height. Similarly, the effective in-plane stress is volume average of the stresses in the layers:

$$\overline{\sigma}_\| = v^{(1)}\sigma_\|^{(1)} + v^{(2)}\sigma_\|^{(2)} + \cdots + v^{(N)}\sigma_\|^{(N)} \tag{6.16}$$

and the out-of-plane stress within each layer is the same and equal to the effective out-of-plane stress:

$$\overline{\sigma}_\perp = \sigma_\perp^{(1)} = \sigma_\perp^{(2)} = \cdots = \sigma_\perp^{(N)} \tag{6.17}$$

Substituting the constitutive Eqs. (6.9, 6.10, 6.11, 6.12 and 6.13) and following the procedure introduced in Ref. [47], we can then solve explicitly for the components of the effective stiffness matrix as:

$$\overline{C}_\perp = \left[\sum_{i=1}^{N} v^{(i)}\left(C_\perp^{(i)}\right)^{-1} \right]^{-1} \tag{6.18}$$

$$\overline{C}_\times = \overline{C}_\perp \left[\sum_{i=1}^{N} v^{(i)}\left(C_\perp^{(i)}\right)^{-1} C_\times^{(i)} \right] \tag{6.19}$$

$$\overline{C}_\times^T = \left[\sum_{i=1}^{N} v^{(i)} C_\times^{(i)T}\left(C_\perp^{(i)}\right)^{-1} \right] \overline{C}_\perp \tag{6.20}$$

$$\overline{C}_\| = \sum_{i=1}^{N} v^{(i)} C_\| + \sum_{i=1}^{N} v^{(i)} C_\times^{(i)T}\left(C_\perp^{(i)}\right)^{-1}\left(\overline{C}_\times - C_\times^{(i)}\right) \tag{6.21}$$

So far, we have been able to derive the effective stiffness components for the macro-scale homogenized element. For an explicit simulation of a battery crush event, an incremental strain tensor is calculated based on the loading and boundary conditions, which we then use to calculate the effective stress and strain histories. Implementing the effective stress and strain, and the constitutive Eqs. (6.9 and 6.10) into Eqs. (6.11–6.21), we can then calculate the stress and strain histories for each individual layer.

There are several advantages to this approach:

(i) The deformation histories for the components can be coupled with the electrical-thermal properties. For example, we can incorporate evolution of conductivity and porosity with layer thickness/volume.
(ii) Specific failure criteria can be defined for each component based on their mechanical responses, with which we can then predict changes in short resistance during the deformation process.
(iii) This homogenization approach can be extended to a nonlinear material model by including components for plastic deformation into the constitutive Eq.

(6.10), while the assumptions behind Eqs. (6.14, 6.15, 6.16 and 6.17) are still applicable.

6.3 Coupling with Electrochemical and Thermal Models

We need a framework to analyze interactions between mechanical deformations and the respective electrochemical and thermal (ECT) processes that take place within the electrochemical cell. An obvious solution is to perform the mechanical simulation first and use the deformed geometry to perform the ECT simulations [48]. This one-way coupling approach is highly efficient for instances where the mechanical event has a very small time constant and can be effectively decoupled from the ECT simulations. The downside is that the model cannot account for continual deformation in geometry nor incorporate response from the ECT simulation results into the mechanical response.

6.3.1 Electrochemical Modeling Framework

The homogenization technique outlined in the previous section was extended to couple the electrochemical, thermal, and electrical models [49]. Effective mechanical properties for the jellyroll were computed using the approach outlined in previous sections. For the electrochemistry, we implemented a pseudo-two-dimensional model [50–52]. In the past, we have also incorporated alternate implementations from [53–55].The details of the model equations and parameters are provided in [52]. When the geometry deforms, volume fractions, porosity, and pore diameters of the different phases change simultaneously. These changes are difficult to decouple experimentally.

Based on the assumptions behind the compression loading conditions, changes in porosity vary linearly with changes in thickness and are thus used to reflect changes in transport properties such as the electrolyte diffusivity, with the loading condition. We do not explicitly model changes to the Bruggeman coefficient. These must be evaluated on a case-by-case basis under complex loading conditions. However, the temporal changes in porosity are captured by the mechanical simulations and are accounted for in our implementation of the electrochemical model.

6.3.2 Numerical Implementation

LS-DYNA has separate FORTRAN subroutines for updating mechanical (umat), thermal (thumat), and element formulation (usrsld, usrshl) variables [42]. For illustrative purposes, we implemented using these on solid elements. The elec-

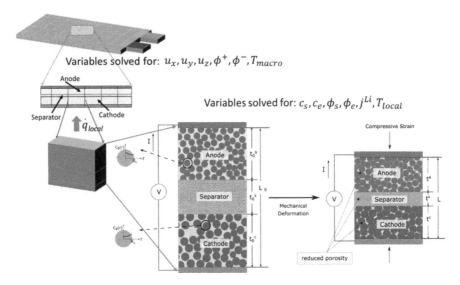

Fig. 6.7 Multiscale approach for modeling lithium-ion cell during mechanical deformation. Reduction in thickness of each layer in turn affects porosity and effective transport properties. Simultaneous coupling of mechanical, electrochemical, and thermal models captures the evolution of electrochemical behavior during deformation. (Figure reproduced with permission from Ref. [49])

trochemical model is solved in a separate geometry (see Fig. 6.7) on each solid mechanical element. Note that this approach allows us to specify a separate mesh and/or time-stepping scheme for the ECT models independent of the mechanical models, although this is not a mandatory requirement. Through thickness strain components read in by the electrochemical subroutine are used to update the corresponding dimensions.

Using a multiscale modeling approach, the mechanics have been modeled using a large deformation macro model. Changes in the macro-scale temperature (T_{macro}) are controlled by a lumped heat source ($\dot{q}_{\text{local}}$) from each individual mesh element which has an electro-thermal model embedded in it. The heat equation (Eq. 6.28) governs heat transfer between elements through thermal conduction. The lumped heat source ($\dot{q}_{\text{local}}$) is a sum of short-circuit heat, joule heat, and reaction heat. Local temperature (T_{local} in Fig. 6.7) is computed as a function of the thickness:

$$\frac{\partial \left(\overline{\rho c_p} T_{\text{macro}}\right)}{\partial t} = \nabla \bullet (\overline{\kappa} \nabla T_{\text{macro}}) + \dot{q}_{\text{local}} \tag{6.28}$$

The mechanical model provides the values of compressive strain for individual electrodes and separator at each time step which is used to update the instantaneous layer thicknesses in the electrochemical model. The thickness t of each layer *i* at the current time step is given by:

$$\frac{t^i}{t^i_0} = \exp\left(\epsilon'^i_{zz}\right) = 1 + \epsilon^i_{zz} \tag{6.29}$$

$\epsilon^i_{zz} = \left(\frac{\Delta L}{L_0}\right)$ is the through thickness component of the engineering strain for layer i. The reference state is denoted with subscript 0 for porosities and thicknesses of each layer i. A change in thickness results in a change in porosity within the components. Effective electrolyte conductivity and diffusivity parameters, which are functions of local porosity, are automatically updated as porosity changes according to Eq. (6.29). Dependence of porosity (ε) on thickness is evaluated using the following expression:

$$\varepsilon^i = \varepsilon^i_0 \left(\frac{t^i_0}{t^i}\right) \tag{6.30}$$

The heat generated by the battery model is passed to the thermal subroutine as a source term. Time scaling is applied to the thermal solver using the parameter TSF (thermal speedup factor) ($=10^3$) under the *CONTROL_THERMAL_SOLVER keyword in LS-DYNA; a similar pseudo-time is used in the mechanical simulations to avoid expensive computational steps when using the explicit solver. The electrochemical model is implemented as a user-defined function call within the mechanical subroutine and is solved once after several mechanical time steps for computational efficiency.

6.3.3 Geometry and Meshing

Meshing of 3D geometries was usually performed using Hypermesh. When defeaturing the computational geometry to eliminate elements that were of poor aspect ratio, implications of these modifications on the outcome of subsequent thermal simulations were used to verify that the simplifications were reasonable. Sensitivity analysis of the results to the mesh quality was routinely performed to address numerical artifacts. Mesh and element types were chosen commensurate with the physics being solved for. The choice of tools used to solve for the mechanical and ECT response also allows for mixing-and-matching of element types. In some instances, the mechanical models were efficiently implemented using shell elements, whereas the CFD simulations associated with the ECT models were performed on solid elements. Within the cell the contiguous electrically conductive regions (e.g., jellyroll, tabs, and terminals) were discretized with a conformal mesh containing shared nodes. Contact was defined between the remaining surfaces using *CONTACT_AUTOMATIC_SINGLE_SURFACE option in LS-DYNA. For the models with one-way coupling, once the mechanical simulations were complete, the deformed mesh was imported into ANSYS Fluent® for the electrochemical-thermal simulations, along with the strain field as a function of geometric coordinates. Nonconforming mesh interfaces were manually defined between each of surface pairs in contact with each other on the deformed geometry. Numerical schemes to stabilize simultaneous simulations (with two-way coupling) are discussed later in the chapter.

6.4 Description of an Electrical Short Circuit

One of the key aspects of simulating the electrical outcome and subsequent thermal response of a battery following mechanical impact is incorporating the onset of electrical short circuits. Simulating an external short circuit, both at the single-cell and at the module or larger scales, is relatively straightforward: imposing a boundary condition where the terminals are separated by a resistor and specifying a discharge current, based on the value of the shunt resistance, are two commonly reported [56, 57] approaches. Implementing an internal short-circuit model is more complicated. Several approaches have been proposed in the past, and the literature on this topic is sparse. In this section, we review a few of these approaches.

6.4.1 Internal Short Circuits with a Fixed Geometry

Under this approach, the internal short circuit is represented by a model domain with a high electronic conductivity, specified as part of the initial geometry (see Fig. 6.8). This model domain usually has a regular geometry (e.g., a cylinder of prescribed radius that connects parts of the positive electrode stack with those of the negative). One common example is simulation of a nail penetration test, where a conductive nail provides a conduit for short circuit within the cell. The model usually imposes a voltage continuity criterion between the nail and the adjacent electrode and/or current collector layer(s). The individual electrodes are initialized to set values of electrode and electrolyte potential consistent with their states of charge and relaxation history prior to the onset of the short circuit. These simulations are insightful in analyzing the trade-offs between the short-circuit resistance and internal impedance of the cell. When the short-circuit resistance is relatively high, or the short-circuit area as a percentage of the electrode's geometric is small, thermal events tend to be contained resulting in a localized short. Thus, a short-circuit event with the same nail resistance has very different outcomes depending on the size of the cell being simulated. These models are also good at identifying secondary heating zones (usually around the cathode tabs) due to much higher current densities flowing through the same tab area under short circuit, compared to normal discharge simulations.

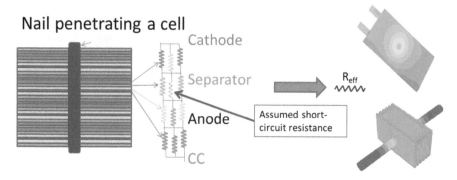

Fig. 6.8 Approximation of a short circuit using a single fixed value resistor is a simple but effective way to simulate hard-shorts (e.g., a low-resistance nail penetration) in both single-cell and multi-cell models

6.4.2   Internal Short Circuits with Evolving Geometries

Short circuits that evolve gradually over time, or under slow mechanical deformation of the cell components, are likely to have the contact area across the conductive components change with time. The framework outlined earlier can be extended to account for these changes. We described mechanical failure criteria for the different cell components in Sect. 6.2.1.4. In coupled mechanical-ECT simulations, these criteria form the basis set for predicting the onset of a short circuit.

The contact area across the short is calculated as the summation of the in-plane area $(a_{s,i})$ over all the elements "i" that have met the failure criterion:

$$A_{\text{Short}} = \sum_{i=1}^{n} a_{s,i} = \sum_{i=1}^{n} L_x^i \left(1 + \epsilon_{xx}^i\right) L_y^i \left(1 + \epsilon_{yy}^i\right) \tag{6.31}$$

Note that the contact area can be updated at each time step by evaluating each element against the failure criterion. The next step is the calculation of the resistance across the short circuit. The electronic conductivity across a short circuit path is calculated from the conductivity of the different elements subjected to short circuit:

$$\frac{1}{\kappa_s a_s} = \sum_{i=1}^{n} \frac{1}{\kappa_{s,i} a_{s,i}} \tag{6.32}$$

Other formulations include using an average of the conductivity values across the short and/or introduction of an interfacial resistance that can be calibrated out using experimental data. Any of these methods can be used with the approach presented here. The short-circuit resistance is then computed as follows:

$$R = \frac{1}{2\kappa_s r} \tag{6.33}$$

where r is the equivalent radius of the element computed using $a_s = \pi r^2$. The value of R obtained from Eq. (6.8) is a spatially distributed function which can be exported into an ASCII format file as an input for the ECT simulations.

Mechanical failure of the elements is one of the criteria to diagnose onset of failure. We need to accurately populate the model with changes in electrical and thermal properties that accompany the onset of mechanical failure. For instance, we can define a drop in electrical resistance prior to the onset of physical contact across the conductive layers, as a function of the local strain in the through-plane direction. Similarly, failure criteria can be defined as a function of the local temperature (e.g., to account for melting or shrinkage in the separator layers). There are also instances where secondary short circuits emerge due to excessive heat generated from the initial mechanically induced short circuit. Specifying thermal criteria for element failure also accounts for these cases.

6.4.3   Stochastic Representation of Short-Circuit Resistances

Short-circuit resistance can be measured as a function of the loading condition or thickness of the cell components or the jellyroll and used as a look-up table in cell-level simulations. However, a persistent challenge with any mechanical impact experiment is the extent of variability in the test outcomes. Even in the most controlled set of experiments, there is a range of values measured for resistances and failure thresholds. The objective is to either minimize these variabilities to within tolerances below which their effect on the safety outcome of the cell does not change significantly. For instance, if the models can be used to show that the EUCAR response will not be different as long as the short resistance for a given scenario does not alter beyond ±30%, such variability can be classified as acceptable.

To perform these simulations, we represent the short circuit as a network of resistances (Fig. 6.9) that include components from the two layers undergoing the short circuit, any barrier layers isolating them, as well as any foreign object debris present within the cell. Each of these resistances can be populated using a series of measurements made at the component level. The contact resistances can be computed as the difference between resistance measured across a two-layer composite and that of the individual components. Note that the contact resistances will also be collection of datapoints, instead of a point value. Once these results are compiled into a look-up table (or a bell curve), we perform stochastic sampling (e.g., using a Latin hypercube scheme) of the individual resistances for each of the network components. The effective short-circuit resistance (R) is then estimated as the most-likely value from a statistically significant set of simulations.

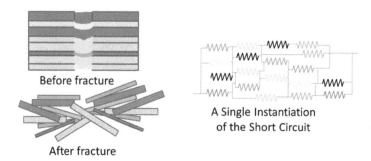

Fig. 6.9 Stochastic representations of short circuits employ experimentally measured values for the electronic conductivities of the cell components. Changes in the thickness of the individual layers account for variations in cell internal resistance before mechanical failure of the individual layers. Short circuit after mechanical failure results in different layers coming into contact which is simulated by a series of circuit representations, each with different probabilities. By computing the effective resistance of these circuits for each instantiation, a distribution of short-circuit resistances and the maximum likelihood is estimated

6.4.4   Electrochemical-Thermal Modeling of Short-Circuit Response

Once the short-circuit resistance has been determined using one of the approaches outlined above, or measured experimentally, the next step is to compute the short-circuit current density and resulting heat generation rates. To account for the effect of porosity of the different layers subjected to short-circuit, on the contact area, we implement the following relationship between the volumetric contact resistance (r_c) and the short circuit resistance (R):

$$r_c = R A_{\text{Short}} \tag{6.34}$$

The short-circuit current density is calculated as part of the ECT model as follows:

$$j_{\text{Short}} = \frac{a \, (\phi_+ - \phi_-)}{r_c} \tag{6.35}$$

where ϕ_j is the potential at the surface of the solid phase j subjected to the short circuit and a is the volume averaged specific area in m^2/m^3:

$$a = \frac{A_{\text{Short}} \left(1 - \varepsilon_j\right)}{t^j a_S} \tag{6.36}$$

The rate of heat generation due to the short circuit event is then given by: [34].

$$\dot{q}_{\text{Short}} = \frac{(j_{\text{Short}})^2 r_c}{a} \tag{6.37}$$

The deformed mesh is exported using the NASTRAN bulk data file (bdf) format, from the mechanical simulations, which is then imported into the ECT model. The resistance values as a function of spatial coordinates at the end of each time step were exported as ASCII files and then mapped onto the deformed mesh through a user-defined function. Importing a complicated geometry after mechanical deformation involves an additional intermediate step, during which mesh elements with poor aspect ratios and/or negative cell volumes are eliminated using a meshing tool.

6.5 Single-Cell Simulation Results

This section shows some examples of results obtained from the current modeling approach computing the electrochemical-thermal responses. One way to modify the short-circuit resistance in the experiments is by varying the size of the mechanical indenter. The variation in the stress-strain response and the corresponding short-circuit resistance for different punch sizes is discussed in detail in [34]. To study the interactions between single-cell response and heat transfer away from the cell, we vary the cooling rates by controlling heat transfer coefficients at the surface of the current collectors.

6.5.1 Simultaneous versus Sequential Simulations

In Sect. 6.2, we discussed solving the mechanical deformation separately, followed by ECT simulations on the deformed mesh. Similarly, we outlined a simultaneously coupled mechanical-ECT (MECT) formulation for cases where the time constants for the different physical phenomena are comparable. Here we show an example emphasizing the difference between the two approaches. Figure 6.10 shows a comparison of the current distribution as arrow plots on the deformed geometry, under identical mechanical loading simulated using the sequential versus simultaneous simulations. As noted, the electrical pathways for the short-circuit currents are captured in a more realistic fashion using the MECT approach. These results also have implications for buildup of thermal resistances and temperature rise.

As one example, depending on the threshold for mechanical failure, the type of damage induced within the cell can result in short circuit across different electronically conductive layers. Accordingly, the magnitude of the short-circuit currents, as well as the resultant temperatures, is strongly dependent on which of the conductive layers within the cell induce the short circuit. For instance, the short circuit resulting from the active material in the cathode coming in contact with the anode coating usually has considerably high resistance to the flow of electrons compared to the short circuit across the two metal current collectors.

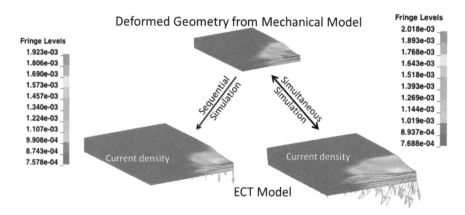

Fig. 6.10 Comparison of current density distribution in the region surrounding the short circuit: for a given loading condition and when the ECT model parameters are identical, the simultaneous coupling of mechanical simulations with the ECT simulations results in accurate representation of the current densities, including direction of the current density vectors

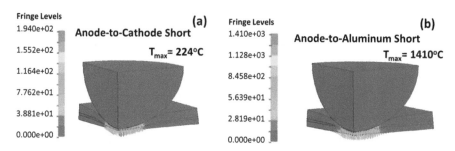

Fig. 6.11 Comparison of the temperature rise from different types of layers from within the cell coming into mechanical contact causing the short-circuit: when the active material from the two electrodes come into contact (**a**), the resultant short-circuit has a relatively higher electrical resistance, and thus the short-circuit current that flows across the different layers is lower compared to that for a short-circuit between the highly conductive carbon anode and the aluminum current collector (**b**). Proportionately, the maximum local temperatures are also different. We showed earlier [58] that a short circuit across the two current collectors, although is of comparable electrical resistance to (**b**) above, results in lower T_{max} because of the better thermal properties of the copper current collectors that aid in dissipation of the short-circuit heats

These effects are captured by the MECT modeling approach outlined above, as shown in Fig. 6.11. These simulations have practical implications for the design of cell components which have robust mechanical, thermal, and electrical failure criteria that must be matched to desired operating and safety constraints.

6.5.2   Species Distribution in the Solid and Liquid Phases

At the onset of short circuit, the active material is depleted much faster around the shorted region. The concentration of active material is more uniform in the through thickness direction near the shorted region due to (i) high current densities passing through the short circuit, (ii) the poor aspect ratio (thickness of the electrodes is far less than the in-plane dimensions), and (iii) higher temperatures facilitating faster kinetics. The gradient in concentration across the electrode as we move away from the short circuit varies with the temperature profiles.

Discharge of cell is often limited by the Li^+ ion transport through the liquid electrolyte. Therefore, understanding electrolyte response during internal short-circuit conditions is crucial. Fig. 6.12 describes the distribution of Li^+ in the electrolyte. Li^+ flux in liquid electrolyte along through thickness direction is given by:

$$\vec{j_{c_e}} = -D_{c_e}{}^{eff}\nabla c_e \tag{6.38}$$

Initially, it can be observed that Li^+ depletion occurs from the active material much faster near the cathode current collector-short-region interface (Fig. 6.12a). This is a result of faster Li^+ diffusion into the active material and resultant current densities at that interface enhanced by the high temperatures as well as proximity to the available electrons. A similar phenomenon is observed near the interface between the copper foil and the short. Subsequently (Fig. 6.12b), Li^+ flux becomes relatively more uniform. For this set of parameters, nonuniformities in electrolyte concentration persist for long durations of time away from the short, with an effect of slow heat up of the cell.

6.5.3   Voltage and Temperature Response Versus Short-Circuit Resistance

Typical cell-level response during a short circuit for various values of resistance is shown in Fig. 6.13. Voltage evolution during short circuit is shown in Fig. 6.13a. Lower values of short-circuit resistance results in a larger voltage drop due to a higher discharging current as expected. There is also a gradual increase in voltage recovery with decreasing resistance at higher values of resistance. The point at which the voltage recovery is maximum coincides with the maximum heat generation rate (Fig. 6.13c). Beyond that, there is a reduction in voltage recovery with further decrease in short resistance because the thermodynamic inability of the cathode to hold back the Li^+ outweighs improvements to transport properties at higher temperatures.

In these models, voltage recovery is clearly related to thermal conditions: Fig. 6.13b shows the time evolution of cell temperature for different values of short-

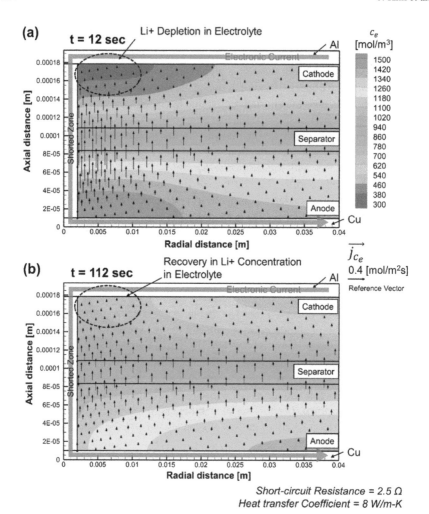

Fig. 6.12 Snapshots of Li$^+$ concentration in electrolyte and diffusion flux of Li$^+$ in electrolyte at different time steps as the short circuit evolves: for a relatively large short-circuit resistance (2.5 Ω) to compensate for the flow of electrons (which happens instantaneously), Li$^+$ migrates over a period of several seconds (**a**) from the anode to the cathode resulting in changes in concentration (contours) and the diffusion flux (arrows). These changes level off with time (**b**). (Figure reproduced with permission from Ref. [59])

circuit resistances. It can be observed that initially, peak temperature increases as the short resistance is lowered. The maximum temperature rise (to 365 K) occurs at $R_{SC} = 1.25$ Ω. Any further decrease in short-circuit resistance results in lower temperature rise. A thermal budget analysis comparing heat generation rates as a function of short-circuit resistances (at 90% SOC or 4 V) shows that this trend is due to decrease in joule heating below 1.25 Ω for the short resistance. As the local heat generation at the short region decreases with decrease in short resistance, most

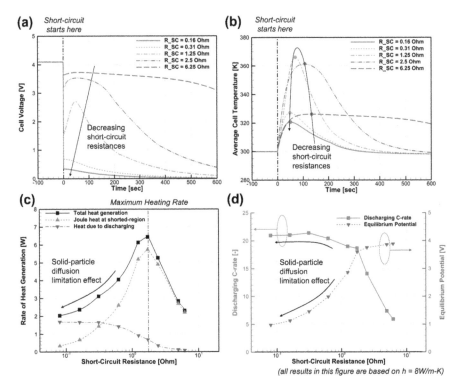

Fig. 6.13 (**a**) Voltage response at various short-circuit resistances. (**b**) Temperature response at various short-circuit resistances. (**c**) Budget analysis for various heat generation mechanism at various short-circuit resistances. (**d**) Discharging rate and equilibrium potential at various short-circuit resistances (heat transfer coefficient of 8 $W \cdot m^{-1} K^{-1}$ is used; data at 90% SOC are selected for Fig. 6.13c–d). (Figure reproduced with permission from Ref. [59])

of the heat contribution comes from cell discharge, causing the cell to move from "local heating" to a "global heating" regime. Similar trend has been reported in Ref. [56, 60, 61]. This trend also helps explain the relative variation in electrolyte concentrations as a function of distance from the short as shown in Fig. 6.12. Li^+ availability thus determines the transition point of a localized short circuit to global heating of the cell. In high power cells, with no significant transport limitations, the concentration profiles are much more uniform and thus result in a global short. These cells generate more heat overall. In high-energy cells with transport limitations (similar to Fig. 6.12), the short circuit lasts for longer durations even though the maximum heat generation rate is lower.

For these set of electrochemical-thermal parameters, the cell internal heat generation at low short-circuit resistances plateaus off due to diffusion limitations within the solid particles. Figure 6.13d shows both equilibrium potential and rate of discharge (both taken at 90% SOC) as a function of short-circuit resistance. With a decrease in short-circuit resistance, the rate of discharge increases until ~20C, and any further decrease in resistance does not lead to an increase in discharge rate. Also, the effective cell equilibrium potential gradually decreases and levels off at roughly 1.0 V as the short-circuit resistance decreases.

6.6   Multi-cell Simulations

The framework for simulating the MECT response of larger format test articles is essentially the same; however, there are complications from aspect ratios and the large number of elements. The additional mechanical constraints, thermal mass from the support fixtures, and propagation of failure from one cell to the others also present additional scenarios to simulate battery failure under practical configurations.

6.6.1   Coupled MECT Simulation of Multi-cell Failure Response

Figure 6.14 shows the geometry of the battery module and the impactor. The module studied in this work consists of 20 cells connected by a bus bar in parallel

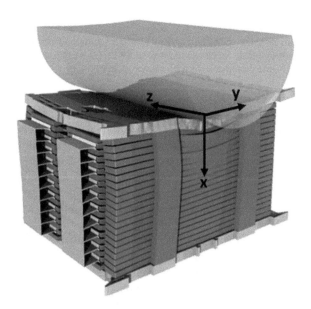

Fig. 6.14 A 20-cell module comprised of 15 Ah pouch format cells subject to mechanical indentation along the x-direction

configuration. The cells are of pouch format with in-plane dimensions of 224 mm ×
164 mm and a thickness of 5.4 mm. The cell charge capacity is 15 Ah with a nominal
voltage 3.65 V, and the cells are charged to 4.15 V prior to each of the tests. Each cell
contains 16 cathode layers and 17 anode layers, stacked periodically and electrically
isolated by polymeric separator layers. A representative-sandwich (RS) model has
been employed to model layer properties efficiently. The parallel connections allow
for multiple cells to discharge simultaneously if one of the cells has an internal short
circuit initiated inside of it, resulting from a flow of current through the bus bar. The
finite element model for the battery module has over 700,000 elements.

An initial velocity of 6.3 m/s was specified for the impact simulations. The back
panel on the battery module is held fixed. Penetration across layers was eliminated
by suitably defining contacts. The jellyroll is treated as a homogeneous solid using
MAT126. Other packaging components were modeled as isotropic-hardening plastic
material. Isotropic electrical properties were defined using EM_MAT001. The case
files were solved on NREL's high-performance computing system, Peregrine [62].
The mechanical time step was set to 1×10^{-8} s, and the EM time step was
1×10^{-5} s. 34 h of computational time was required to simulate 3 ms of the impact
test using 60 large-memory (256 GB) 16-core nodes.

6.6.2 Visualization of Multi-cell Results

In 3D simulation of deforming geometries, especially if they are of large format,
such as an entire battery module, visualization of large datasets encompassing
a variety of physical variables comprehensively can be a challenging problem.
Coupled modeling of mechanical, electrical, and thermal phenomena results in
highly multivariate results across millions of geometric elements. For the different
cell components, electrical or thermal conductivity can vary by several orders of
magnitude. During a mechanical crush test, for example, the current density on
cathode and anode reaches around 300 A/m^2, but at the battery tabs, it can reach
as high as 25,000 A/m^2. To help tease apart this complexity, we resorted to Blender
[63] and ParaView [64], using the geometry and results exported from LS-PrePost,
the default pre- and post-processing interface from LSTC. A custom macro script
can be used to call LS-PrePost for processing multiple datasets. ParaView was used
to generate animations of the surface and interior of the module geometry, generated
from STL files. OSPRay is a CPU-only rendering tool built on Intel's Embree [65]
that provides superscalar accelerated CPU ray tracing. Output from LS-PrePost was
exported to VRML2 format files which were in turn converted to PLY format using
MeshLab [66]. The individual frames within the clips were rendered using a Python
script from ParaView. The results are shown in [67]. The current density distribution
on the active material layers along the direction of mechanical crush is important to
study propagation of internal short circuits from one layer to the other. To visualize
the in-plane current density on the surface of the anodes and cathodes, we produced

an illumination-based visualization using Blender's Cycles rendering system with a surface light emission shader.

Under open circuit conditions, there is no internal in-plane current. However, during an external crush if one or more of the cell components reaches or exceeds the respective mechanical failure criterion, this results in a drastic reduction in electrical resistance. Alternate pathways for the electric current to flow from the positive to the negative electrodes are generated. Of these numerous pathways, the evolution of an electrical short circuit across specific sets of components is determined in this model by i) the physical separation between the conducting layers, ii) the rate of decrease of electrical resistance in the layer across, and iii) the presence of electrical conduction pathways that carry charge away from the element subjected to mechanical failure.

Simultaneously visualizing multiple field variables such as the components of the von Mises stress together with the local current density distribution allows us to accurately determine the location for onset of mechanical failure or electrical short circuit. Tracking these failure points over time provides insights on the mechanism of failure propagation. For instance, in Fig. 6.15a, b, we see that the structural integrity of the end plates significantly reduces damage to the cells' internal components; however, from Fig. 6.16, we observe that electrical short circuits happening in the cells farthest from the impactor are caused by the mechanical resistance of the side brackets located right next to these cells. Such results have design implications for battery packs and provide metrics for spacing between cells and/or mechanical properties of packaging between the end plates and the cells.

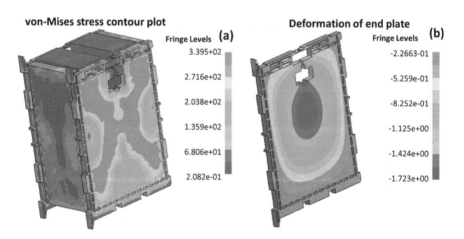

Fig. 6.15 Contours of the von Mises stress (**a**) and deformation on the front plate (**b**) during a frontal impact simulation: the mechanical load was along the x-axis (see Fig. 6.14). The maximum deformation on the end plate was 1.723 mm. In turn, this prevented the layers within the cell from accumulating significant amount of stresses in the through-plane direction

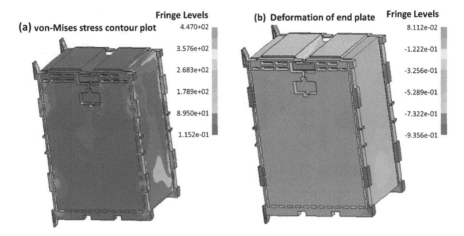

Fig. 6.16 Contours of the von Mises stress (**a**) and deformation on the module packaging (**b**) during a side impact simulation: the mechanical load was along the z-axis (see Fig. 6.14). The maximum deformation on the side brackets was 0.9356 mm. This value is far less compared to that for the frontal impact case shown in Fig. 6.15. This in turn results in several secondary short circuits along the edges. (See Ref. [67])

6.6.3   Design Trade-Offs Using MECT Simulations

Next, we show an example of conducting design analyses using the MECT simulations discussed above. Two cases of mechanical impact on the same battery module were simulated: one case representing a frontal impact and another lateral impact from the thickness direction of the cells (Figs. 6.14, 6.15 and 6.16). In both these simulations, the impact load was maintained the same for a given area; the overall objective is to minimize any mechanical deformation of the cells. For frontal impact, a large section of the end plate shows a higher von Mises stress and subsequent deformation. The lateral impact, although affecting a smaller section of the packaging, shows a higher value for the stress the side brackets are subject to. Another difference to note, compared to the frontal impact, is that the side brackets accumulate higher stresses in favor of lower deformation in the in-plane direction for the cells. This is critical to prevent short circuits along the edges of the cells. These calculations are helpful in determining the thickness and material choices for the end plates versus the side brackets.

In this study, we compare different cooling-channel designs for these two cases as shown in Fig. 6.17. Two sets of simulations were carried out: in the first set, the loading conditions were maintained as the same from the previous case studies (frontal versus lateral impact) for a continuous channel cooling loop. In this instance, the side impact resulted in localized damage to the cooling channel, and because there are no redundancies in this design, the flow was constricted due to the mechanical deformation, resulting in a higher local temperature, compared to the front impact case. We explored an alternate design where multiple cooling loops in

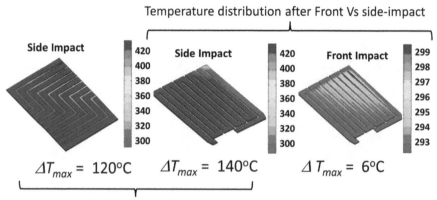

Fig. 6.17 Comparison of the response of different cooling-channel designs to mechanical impact across multiple orientations: localized damage to cooling channels results in high temperature rise which can be mitigated by redundant channel designs. The packaging design must be resilient in addressing multiple failure modes

parallel were used instead of a single loop; in this case, the temperature distribution across the cell was more uniform for identical loading conditions.

6.7 Other Factors Influencing Battery Safety

The sections above outline in detail the test conditions as well as simulation procedure we used to study mechanical deformation in lithium-ion cells and subsequent evolution of electrochemical states and temperature rise. These results show that there are several factors that influence the safety outcome of a given cell. In this section, we outline a few other factors that are commonly considered when assessing safety outcomes of batteries under practical scenario.

6.7.1 Strain-Rate Effects

Most of the results reported thus far, correspond to material properties for the mechanical models that were extracted from quasi-steady-state or low strain-rate (<1 s^{-1}) measurements. Within this range of strain rates, there is not a significant difference in the mechanical properties of interest (see Chap. 5) and, in turn, the outcome of the ECT simulations. However, when we conducted high strain-rate deformation tests using a Hopkinson's bar-type setup, the mechanical response of the electrodes was different from those reported earlier. Higher ultimate strengths

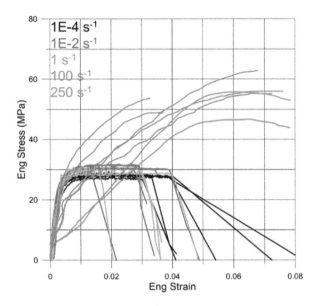

Fig. 6.18 Strain-rate effects tend to be less pronounced when scanning over a limited range of strains (e.g., all quasi-steady-state tensile measurements for cathodes showed a maximum engineering stress of around 30 MPa or less) whereas under dynamic loading conditions, the ultimate stress was much higher and exhibited a significantly different dependence on strain rate

and failure strains were observed for strain rates upward of $100\,\mathrm{s}^{-1}$ (see Fig. 6.18). These constitutive relationships can be readily included in the modeling framework outlined earlier, by directly using experimental data for the load curves as input to the mechanical simulations in LS-DYNA.

6.7.2   Temperature Effects

Arrhenius-type relationship showing temperature dependence for ECT parameters such as diffusivities and conductivities has been widely reported in the literature (see, e.g., Gu and Wang [52]). We carried out several sets of measurements that account for temperature dependence of mechanical properties. For changes in ambient temperature at which the batteries are operated (e.g., 0-60 °C), linear extrapolation of the mechanical properties measured at 25 °C were reasonable [68]. Of particular interest are changes in mechanical properties at temperatures closer to the thermal runaway reaction conditions. Our initial results show an exponential weakening of the coating strength with drastic increase in sample temperatures upward of 120 °C (see Fig. 6.19). However, there is still a gap in the literature in this area, quantifying the effect of abuse reaction temperatures on the constitutive relationships.

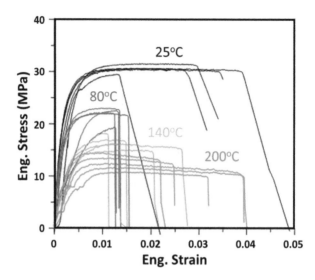

Fig. 6.19 Most datasets in the literature tend to focus on temperature dependence of mechanical (and ECT) properties within the nominal operating range of the cell. Mechanical properties at temperatures close to the onset of thermal runaway reactions have a pronounced effect on the safety response of cell components. This figure shows that the tensile strength of cathodes tends to reduce by over 60% at 200 °C compared to the values measured at 25 °C

6.7.3 Aging Effects

The influence of cycling on degradation of electrochemical processes has been documented widely in the literature [69, 70]. Changes in thermal properties between wet and dry electrodes and [71] influence of contact pressure [72, 73] have also been studied. Changes to mechanical properties, and how these influence safety outcomes, are still an active area of research [19, 74]. We have measured mechanical properties of electrodes subject to the same aging conditions, but different storage durations subsequently. Figure 6.20 shows some data from cycling studies that were conducted over a period of 10 years. Other groups have reported changes to mechanical properties, from cells subjected to different cycling conditions (C-rates, voltage windows, etc.). As noted earlier, these changes can be incorporated into the MECT models by using the appropriate loading curves for the constitutive relationships.

Aging of the cells also alters thermal properties of the cell components. Figure 6.21 discusses one example of these changes. We measured thermal conductivity on electrode samples collected from the 40 Ah pouch cells subjected to aging: the bulk conductivity was measured per ASTM Standard D5470–12, and the interfacial conductivity was deduced by measuring the effective conductivity in a stack of different layers and subtracting the baseline bulk contributions. Initially, the thermal conductivity of the cathode is the lowest and is a significant contributor

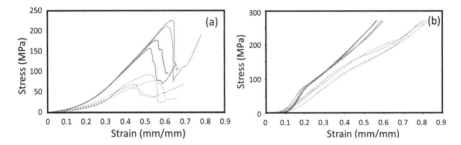

Fig. 6.20 Changes in compression strengths of (**a**) graphite anodes and (**b**) NMC-111 cathodes: the solid lines are loading curves measured under compression using samples harvested from different layers of a 40 Ah pouch cell at the beginning of life, and the dotted lines correspond to similar measurements after the cells were cycled between 3.0 and 4.1 V at 25 °C. (Figure modified with permission from Ref. [74])

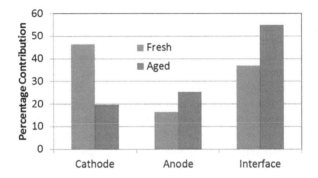

Fig. 6.21 Changes to thermal impedance of the anode versus the cathode and the contact impedances were quantified using samples obtained from the aging studies outlined in Fig. 6.20. With cell aging, the interfacial thermal impedances become a significant source of heat generation within the cell. The cathode's thermal contribution, while does not decrease in absolute value, becomes relatively less important: as the electrical resistance at the anode surface builds up, the thermal impedance in the bulk of the anode and at the interface dominated the thermal response of the cell under test

to the thermal signature of the cell. With aging, we measured an increase in the electrochemical impedance (EIS) of the anode which corroborated well with the buildup of the thermal contributions from the anode and the corresponding interface.

6.7.4   State-of-Charge Effects

Thus far, there has been one study [17] that captures trends in mechanical properties in 18650 format cells as a function of lithium content within the electrodes; however, in our in-house measurements, the effect of state of charge on mechanical properties were within experimental error. Temperature effects at higher states of charge must be accounted for separately, in our models.

Other conditions such as nonuniform distribution of the reaction stresses due to cycling at higher C-rates or thermal shock can be simulated using the modeling framework outlined in this chapter, using the appropriate boundary conditions for the ECT simulations, alongside temperature-dependent mechanical properties.

6.8 Numerical Considerations

Numerical convergence can be an issue when performing coupled multi-physics simulations across different length and/or timescales. We have implemented the framework outlined in this chapter, with independent control of time steps for mechanics versus the ECT simulations. Similarly, we have refined the mesh used for these simulations to different degrees [45, 49]. Other groups have implemented unconditionally stable solvers [75]. Some of these approaches involve linearizing select source terms [76]. Others have resorted to modified kinetic expressions that set bounds on the local current density based on mass transfer limitations, under high-rate discharge [77]. More recently, we reported the use of an analytical Jacobian wherein select entries are modified to impose physical limits on the local electrolyte concentration being solved for [59]. A similar approach can be developed for other internal states within the cell [78]. Such modifications were found to dramatically enhance stability of the simulations in our studies.

Other considerations include careful development of material balances in systems involving volume changes [53]. For simple geometries, variable transformations offer reliable alternatives; but as we scale these simulations to complex module or pack-level geometries, efficient ways to define element deletion criteria [34] to eliminate mesh elements with poor aspect ratios and developing rigorous species conservation equations [79] become essential. Numerical trade-offs between accurately capturing failure patterns versus obtaining physically meaningful temperature evolution profiles, within a reasonable amount of computational time, are still handled on a case-by-case basis by the battery modeling community.

6.9 Next Steps

Simulation of safety outcomes for vehicle batteries has come a long way during the last 10 years, from designing fail-safe enclosures for batteries to understanding abuse reaction mechanisms and key parameters that control propagation of failure from a single cell to adjacent modules. There are a few gaps in developing and parameterizing these models that we have called out in previous sections. There are emerging experimental capabilities that allow for elaborate measurement of internal states of batteries, at different scales. Access to computational power has become a lot more prevalent, allowing for detailed simulation of physical phenomena. In this section, we call out a few areas where these tools are continuing to improve our understanding of battery safety.

6.9.1 Gas-Phase Reactions

Pressure buildup within the cells and subsequent mechanical venting of the container has long been a topic area of interest, both from a design and from a safety mitigation perspective. Design of vents at which there is a controlled release of pressure has traditionally been carried out by determining the failure threshold that does not interfere with nominal cell operation. Detailed gas-generation [80, 81] and gas-phase reaction mechanisms [82] that account for buildup of shock waves resulting in secondary mechanical damage within the module are under development. These phenomena have long been reported in the experimental literature; but accurate representation in a mathematical framework has been challenging due to lack of reproducible datasets. Similarly, several experimental results have reported [83–85] an initial cell-venting stage where the solvent vapors are released. This event is often followed by a much more reactive vent that results from reaction of the cell ejecta from the first vent with ambient oxygen. We recently showed [86] that there is a careful trade-off between excessive buildup of pressure before venting of the can and release of ejecta at a lower pressure threshold, at which the composition of the gaseous effluents can be managed to control any secondary reactions in the gas phase. Condensates from the solvent have been observed on cells not in the immediate vicinity of the cell undergoing a runaway event. Detailed fault-tree analyses that govern failure modes of individual cells and subsequent propagation mechanisms have been documented in detail [87]. Incorporating gas-phase reactions alongside flow simulations will likely be important, especially in very large-format battery installations.

6.9.2 Data Science and Battery Safety

Several recent reports have begun to explore pattern identification as a means to identify impending failure. Relatively larger sets of replicates for the different test conditions are reported in the recent literature [88]. Some initiatives on consolidating large databanks have been proposed [89, 90]. These tools are still under development. There is a strong need to identify metrics for error bars on the covariance matrices that help distinguish trends that result from physical degradation against variability in experimental data. Data quality has persistently been a challenge with safety simulations, in addition to the limited quantities in which good-quality data has been reported. Developing methodology to filter background noise in the data has been another area of emphasis that development of physics-based models will benefit. At least for the foreseeable future, we do not anticipate extensive datasets (e.g., similar to cell aging databanks that span several decades and thousands of cycles in a variety of cell formats and chemistries) becoming available on battery safety. These limitations are being addressed by (a) significantly enhancing the instrumentation on the limited number of experiments conducted and (b) resorting to statistical tools that account for these limitations. Development of fault-trees based on comprehensive analysis of data are currently underway [91]. Arguments on the stochastic nature of the test outcomes are still being refined. Chapter 7 discusses this topic in more detail. With the augmented sensing capabilities and a closer look at experimental variabilities, different research groups are improving reproducibility of safety tests.

6.9.3   Summing It All Up

Given the subjectivity in interpreting safety test results, and the need to arrive at practical design metrics for batteries used in different applications, several groups have developed methods to build response surfaces that compare the relative safety outcomes from simulations. We have approached this objective by conducting a sensitivity analysis across design variables that can be controlled and monitoring the heat generation rates under a variety of different failure scenarios [92]. Plotting rate of heat generation against heat-transfer rates allows for a generic comparison of different cell designs and associated failure scenario on a "safety map" [93, 94]. Figure 6.22 shows an example of such a safety plot.

Criteria for safe design are often determined by system-level specifications. For instance, there are module design specifications that set a maximum local value for temperature as the metric to calibrate safe design. Battery engineers can then classify response from different test articles as (i) "inherently safe region" meaning, no engineering workaround is necessary to meet the specified safety criterion under all operating/failure scenario, and (ii) "operating region engineered to be safe" meaning, there are design aspects that can be modified to ensure that the specified safety criterion is satisfied. Several passive mitigation strategies have been proposed [95]. For instance, the spacing between the cells can be increased to prevent excessive temperature buildup within the module and iii) "unsafe region" meaning there are no design workarounds possible and the mitigation strategy is to avoid cell operation in this region. This approach allows one to determine not just absolute criteria for safe operation, but also evaluate margins for error within the constraints allowed.

The approach outlined in this chapter, to model safety response of batteries, is fairly generic and can be readily extended to different chemistries. Chemistry-specific failure modes and degradation mechanisms have to be identified [96–100], pertinent source terms and constitutive relationships parameterized using experimental data, and relevant target metrics for mitigation can be identified using the broad framework presented here.

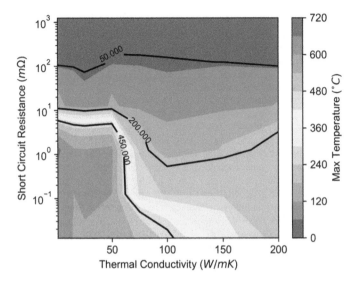

Fig. 6.22 A generic plot comparing the heat generation rate against the heat transfer away from the cell is a powerful tool to compare different design solutions: the heat generation rate can be varied, as is shown in this example, by altering the short-circuit resistance and the response of the cell under different design scenario where the conductivity of the coolant is changed and is monitored by tracking the maximum local temperature. Maximum temperature limits can be specified to determine bounds of safe operation, mitigation strategies, and preemptive design

Acknowledgments This study was supported by Computer Aided Engineering for Batteries (CAEBAT) project of the Vehicle Technologies Office, Office of Energy Efficiency and Renewable Energy, US Department of Energy under contact number WBS1.1.2.406. The research was performed using computational resources sponsored by the Department of Energy's Office of Energy Efficiency and Renewable Energy, located at the National Renewable Energy Laboratory. Contributions made from current and previous members of the Electrochemical Energy Storage Group at NREL are acknowledged.

References

1. Hollmotz L, Hackmann M (2006) Lithium ion batteries for hybrid and electric vehicles – risks, requirements and solutions out of the crash safety point of view. EVS 22. Paper 11-0269, pp 1–9
2. Kermani G, Sahraei E (2017) Review: characterization and modeling of the mechanical properties of lithium-ion batteries. Energies 10(11):1730. https://doi.org/10.3390/en10111730
3. Bartolo M (2012) EV vehicle safety. Electric vehicle safety technical symposium, pp 6–15
4. Doughty DH (2012) Vehicle battery safety roadmap guidance, NREL Report No. SR-5400-54404

5. Deng J, Bae C, Marcicki J, Masias A, Miller T (2018) Safety modelling and testing of lithium-ion batteries in electrified vehicles. Nat Energy 3(4):261–266. https://doi.org/10.1038/s41560-018-0122-3
6. Sahraei E, Meier J, Wierzbicki T (2014) Characterizing and modeling mechanical properties and onset of short circuit for three types of lithium-ion pouch cells. J Power Sources 247:503–516. https://doi.org/10.1016/j.jpowsour.2013.08.056
7. Marcicki J, Zhu M, Bartlett A, Yang XG, Chen Y, Miller I, L'Eplattenier T, Caldichoury P (2017) A simulation framework for battery cell impact safety modeling using LS-DYNA. J Electrochem Soc 164(1):A6440–A6448
8. Zhang C, Santhanagopalan S, Sprague MA, Pesaran AA (2015) A representative-sandwich model for simultaneously coupled mechanical-electrical-thermal simulation of a lithium-ion cell under quasi-static indentation tests. J Power Sources 298:309. https://doi.org/10.1016/j.jpowsour.2015.08.049
9. Cannarella J, Liu X, Leng CZ, Sinko PD, Gor GY, Arnold CB (2014) Mechanical properties of a battery separator under compression and tension. J Electrochem Soc 161(11):F3117–F3122. https://doi.org/10.1149/2.0191411jes
10. Sahraei E, Campbell J, Wierzbicki T (2012) Modeling and short circuit detection of 18650 Li-ion cells under mechanical abuse conditions. J Power Sources 220:360–372. https://doi.org/10.1016/j.jpowsour.2012.07.057
11. Lai W-J, Ali MY, Pan J (2014) Mechanical behavior of representative volume elements of lithium-ion battery modules under various loading conditions. J Power Sources 248:789–808
12. Lai W-J, Ali MY, Pan J (2014) Mechanical behavior of representative volume elements of lithium-ion battery cells under compressive loading conditions. J Power Sources 245:609–623
13. Luo H, Juner Z, Sahraei E, Xia Y (2018) Adhesion strength of the cathode in lithium-ion batteries under combined tension/shear loadings. RSC Adv 8:3996–4005
14. Sheidaei A, Xiao X, Huang X, Hitt J (2011) Mechanical behavior of a battery separator in electrolyte solutions. J Power Sources 196(20):8728–8734
15. Sahraei E, Hill R, Wierzbicki T (2012) Calibration and finite element simulation of pouch lithium-ion batteries for mechanical integrity. J Power Sources 201:307–321. https://doi.org/10.1016/j.jpowsour.2011.10.094
16. Sahraei E, Kahn M, Meier J, Wierzbicki T (2015) Modelling of cracks developed in lithium-ion cells under mechanical loading. RSC Adv 5(98):80369–80380. https://doi.org/10.1039/c5ra17865g
17. Xu J, Liu B, Hu D (2016) State of charge dependent mechanical integrity behavior of 18650 lithium-ion batteries. Sci Rep 6:1–11. https://doi.org/10.1038/srep21829
18. Xu J, Liu B, Wang X, Hu D (2016) Computational model of 18650 lithium-ion battery with coupled strain rate and SOC dependencies. Appl Energy 172:180–189
19. Fink K, Santhanagopalan S, Hartig J, Cao L (2019) Characterization of aged Li-ion battery components for direct recycling process design. J Electrochem Soc 166(15):A3775–A3783. https://doi.org/10.1149/2.0781915jes
20. Steele LAM, Lamb J, Gorsso C, Quintana J, Torres-Castro J, Stanley L (2017) Battery safety testing. In: 2017 vehicle technologies office energy storage annual merit review, p ES203
21. Dixon B, Mason A, Sahraei E (2018) Effects of electrolyte, loading rate and location of indentation on mechanical integrity of li-ion pouch cells. J Power Sources 396:412–420. https://doi.org/10.1016/J.JPOWSOUR.2018.06.042
22. Zhang X, Sahraei E, Wang K (Sep. 2016) Deformation and failure characteristics of four types of lithium-ion battery separators. J Power Sources 327:693–701
23. Luo H, Jiang X, Xia Y, Zhou Q (2015) Fracture mode analysis of lithium-ion battery under mechanical loading. In: ASME 2015 International mechanical engineering congress and exposition. Nov 13, 2015. https://doi.org/10.1115/IMECE2015-52595
24. Sahraei E, Kahn M, Meier J, Wierzbicki T (2015) Modelling of cracks developed in lithium-ion cells under mechanical loading. RSC Adv 5(98):80369–80380. https://doi.org/10.1039/C5RA17865G

25. Sahraei E, Wierzbicki T, Hill R, Luo H (2010) Crash safety of lithium-ion batteries towards development of a computational model. In: SAE technical paper. pp. 2010–01–1078
26. Xia Y, Wierzbicki T, Sahraei E, Zhang X (2014) Damage of cells and battery packs due to ground impact. J Power Sources 267:78–97. https://doi.org/10.1016/j.jpowsour.2014.05.078
27. Avdeev I, Gilaki M (2014) Structural analysis and experimental characterization of cylindrical lithium-ion battery cells subject to lateral impact. J Power Sources 271:382–391. https://doi.org/10.1016/j.jpowsour.2014.08.014
28. Kermani G, Keshavarzi MM, Sahraei E (2021) Deformation of lithium-ion batteries under axial loading: analytical model and representative volume element. Energy Rep 7:2849–2861
29. Liu B et al (2019) Safety issues and mechanisms of lithium-ion battery cell upon mechanical abusive loading: a review. Energy Storage Mater 24:85–112. https://doi.org/10.1016/j.ensm.2019.06.036
30. Zhang C, Santhanagopalan S, Sprague MA, Pesaran AA (2015) A representative-sandwich model for simultaneously coupled mechanical-electrical-thermal simulation of a lithium-ion cell under quasi-static indentation tests. J Power Sources 298:309–321. https://doi.org/10.1016/j.jpowsour.2015.08.049
31. Zhang C, Santhanagopalan S, Sprague MA, Pesaran AA (2015) Coupled mechanical-electrical-thermal modeling for short-circuit prediction in a lithium-ion cell under mechanical abuse. J Power Sources 290:102. https://doi.org/10.1016/j.jpowsour.2015.04.162
32. Newman J, Tiedemann W (1975) Porous electrode theory with battery applications. AICHE J 21(1):25–41
33. C. Zhang, S. Santhanagopalan, A. Pesaran, E. Sharaei, and T. Wierzbicki, Coupling of mechanical behavior of lithium ion cells to electrochemical-thermal models for battery crush, Presented at the Annual Merit Review of the Vehicle Technologies Office, Washington D.C., June 2015
34. Zhang C, Santhanagopalan S, Sprague MA, Pesaran AA (2016) Simultaneously coupled mechanical-electrochemical-thermal simulation of lithium-ion cells. ECS Trans 72(24):9. https://doi.org/10.1149/07224.0009ecst
35. Zhang C, Xu J, Cao L, Wu Z, Santhanagopalan S (2017) Constitutive behavior and progressive mechanical failure of electrodes in lithium-ion batteries. J Power Sources 357:126. https://doi.org/10.1016/j.jpowsour.2017.04.103
36. Sahraei E, Bosco E, Dixon B, Lai B (2016) Microscale failure mechanisms leading to internal short circuit in Li-ion batteries under complex loading scenarios. J Power Sources 319:56–65
37. S. Santhanagopalan, C. Zhang, C. Yang, A. Wu, L. Cao, and A. A. Pesaran, Modeling mechanical failure in lithium-Ion batteries, Presented at the Annual Merit Review of the Vehicle Technologies Office, Washington D.C., June 2017
38. Wang H, Simunovic S, Maleki H, Howard JN, Hallmark JA (2016) Internal configuration of prismatic lithium-ion cells at the onset of mechanically induced short circuit. J Power Sources 306:424–430
39. Sahraei E, Campbell J, Wierzbicki T (Dec. 2012) Modeling and short circuit detection of 18650 Li-ion cells under mechanical abuse conditions. J Power Sources 220:360–372. https://doi.org/10.1016/J.JPOWSOUR.2012.07.057
40. Wierzbicki T, Sahraei E (2013) Homogenized mechanical properties for the jellyroll of cylindrical Lithium-ion cells. J Power Sources 241:467–476
41. J. O. Hallquist, Livermore software technology corporation, LS-DYNA theory manual, 2006
42. Borrvall T, Erhart T (2006) A user-defined element interface in LS-DYNA v971, No 4, pp 25–34
43. Marcicki J et al (2016) Battery abuse case study analysis using LS-DYNA. In: Proceedings of the 14th LS-DYNA user conference, Dearborn, pp 12–14
44. Zhu J, Zhang X, Sahraei E, Wierzbicki T (Dec. 2016) Deformation and failure mechanisms of 18650 battery cells under axial compression. J Power Sources 336:332–340. https://doi.org/10.1016/J.JPOWSOUR.2016.10.064

45. Yin H, Ma S, Li H, Wen G, Santhanagopalan S, Zhang C (2021) Modeling strategy for progressive failure prediction in lithium-ion batteries under mechanical abuse. eTransportation 7:100098. https://doi.org/10.1016/j.etran.2020.100098

46. Coman PT, Darcy EC, Veje CT, White RE (2017) Modelling Li-ion cell thermal runaway triggered by an internal short circuit device using an efficiency factor and Arrhenius formulations. J Electrochem Soc 164(4):A587–A593. https://doi.org/10.1149/2.0341704jes

47. Zhang C, Waksmanski N, Wheeler VM, Pan E, Larsen RE (2015) The effect of photodegradation on effective properties of polymeric thin films: a micromechanical homogenization approach. Int J Eng Sci 94:1–22

48. Zhang C, Santhanagopalan S, Sprague MA, Pesaran AA (2015) Coupled mechanical-electrical-thermal modeling for short-circuit prediction in a lithium-ion cell under mechanical abuse. J Power Sources 290:102–113. https://doi.org/10.1016/j.jpowsour.2015.04.162

49. Mallarapu A, Kim J, Carney K, DuBois P, Santhanagopalan S (2020) Modeling extreme deformations in lithium ion batteries. eTransportation, p 100065. https://doi.org/10.1016/j.etran.2020.100065

50. Smith K, Wang C-Y (2006) Solid-state diffusion limitations on pulse operation of a lithium ion cell for hybrid electric vehicles. J Power Sources 161:628–639. https://doi.org/10.1016/j.jpowsour.2006.03.050

51. Doyle M, Newman J, Gozdz AS, Schmutz CN, Tarascon JM (1996) Comparison of modeling predictions with experimental data from plastic lithium ion cells. J Electrochem Soc 143(6):1890–1903. https://doi.org/10.1149/1.1836921

52. Gu WB, Wang C-Y (2000) Thermal and electrochemical coupled modeling of a lithium-ion cell, in lithium batteries. ECS Proc 99–25(1):748–762

53. Mai W, Colclasure A, Smith K (2019) A reformulation of the pseudo2d battery model coupling large electrochemical-mechanical deformations at particle and electrode levels. J Electrochem Soc 166(8):A1330–A1339. https://doi.org/10.1149/2.0101908jes

54. Kim G-H, Smith K, Lee K-J, Santhanagopalan S, Pesaran A (2011) Multi-domain modeling of lithium-ion batteries encompassing multi-physics in varied length scales. J Electrochem Soc 158(8):A955. https://doi.org/10.1149/1.3597614

55. Ramadass P, Haran B, Gomadam PM, White RE, Popov BN (2004) Development of first principles capacity fade model for Li-ion cells. J Electrochem Soc 151(2):A196. https://doi.org/10.1149/1.1634273

56. Zhao W, Luo G, Wang C-Y (2015) Modeling nail penetration process in large-format Li-ion cells. J Electrochem Soc 162(1):A207–A217. https://doi.org/10.1149/2.1071501jes

57. Zhao W, Luo G, Wang C-Y (2015) Modeling internal shorting process in large-format Li-ion cells. J Electrochem Soc 162(7):A1352–A1364. https://doi.org/10.1149/2.1031507jes

58. Santhanagopalan S, Ramadass P, Zhang JZ (2009) Analysis of internal short-circuit in a lithium ion cell. J Power Sources 194(1):550. https://doi.org/10.1016/j.jpowsour.2009.05.002

59. Kim J, Mallarapu A, Santhanagopalan S (2020) Transport processes in a Li-ion cell during an internal short-circuit. J Electrochem Soc.https://doi.org/10.1149/1945-7111/ab995d

60. Fang W, Ramadass P, Zhang Z (2014) Study of internal short in a Li-ion cell-II. Numerical investigation using a 3D electrochemical-thermal model. J Power Sources 248:1090–1098. https://doi.org/10.1016/j.jpowsour.2013.10.004

61. Kim G-H, Smith K, Pesaran AA (2009) Lithium-ion battery safety study using multi-physics internal short-circuit model

62. NREL High Performance Computing (2015) http://hpc.nrel.gov/users/systems/peregrine

63. Blender (2016) Blender – a 3D modelling and rendering package, 2016. www.blender.org

64. Ayachit U (2015) The ParaView guide: a parallel visualization application, Kitware

65. Wald I, Woop S, Benthin C, Johnson GS, Ernst M (2014) Embree – a kernel framework for efficient CPU ray tracing. ACM Trans Graph, pp 1–8

66. Cignoni P, Callieri M, Corsini M, Dellepiane M (2008) MeshLab: an open-source mesh processing tool. In: European Italian conference, pp 129–136

67. No Title. https://www.youtube.com/watch?v=Hb5JWbcrVEY&feature=youtu.be

68. Santhanagopalan S (2017) Efficient simulation and abuse modeling of mechanical-electrochemical-thermal phenomena in lithium-ion batteries. In: Vehicle Technologies Office, Annual Merit Review, p ES298
69. Zhang Q, White RE (2008) Capacity fade analysis of a lithium ion cell. J Power Sources 179(2):793–798. https://doi.org/10.1016/J.JPOWSOUR.2008.01.028
70. Dai Y, Cai L, White RE (2014) Simulation and analysis of stress in a Li-ion battery with a blended LiMn2O4 and LiNi0.8Co0.15Al 0.05O2 cathode. J Power Sources 247:365–376. https://doi.org/10.1016/j.jpowsour.2013.08.113
71. Maleki H, Al Hallaj S, Selman JR, Dinwiddie RB, Wang H (Mar. 1999) Thermal properties of lithium-ion battery and components. J Electrochem Soc 146(3):947–954. https://doi.org/10.1149/1.1391704
72. Zhang YC, Briat O, Deletage J-Y, Martin C, Gager G, Vinassa J-M (2018) Characterization of external pressure effects on lithium-ion pouch cell. In: 2018 IEEE international conference on industrial technology, pp 2055–2059, https://doi.org/10.1109/ICIT.2018.8352505
73. Mohtat P, Lee S, Siegel JB, Stefanopoulou AG (2021) Reversible and irreversible expansion of lithium-ion batteries under a wide range of stress factors. J Electrochem Soc 168(10):100520. https://doi.org/10.1149/1945-7111/ac2d3e
74. Wu Z, Cao L, Hartig J, Santhanagopalan S (Jul. 2017) (Invited) effect of aging on mechanical properties of lithium ion cell components. ECS Trans 77(11):199–208. https://doi.org/10.1149/07711.0199ecst
75. Pathak M, Sonawane D, Lawder MT, Subramanian V (2015) Robust fail-safe iteration free solvers for battery models. ECS Meet Abstr MA2015-02, 172. https://doi.org/10.1149/ma2015-02/2/172
76. Rodríguez A, Plett GL, Trimboli MS (2017) Fast computation of the electrolyte-concentration transfer function of a lithium-ion cell model. J Power Sources 360:642–645. https://doi.org/10.1016/J.JPOWSOUR.2017.06.025
77. Mao J, Tiedemann W, Newman J (2014) Simulation of temperature rise in Li-ion cells at very high currents. J Power Sources 271:444–454. https://doi.org/10.1016/j.jpowsour.2014.08.033
78. Kim J, Mallarapu A, Santhanagopalan S, Newman J (2023) Efficient numerical treatment for solid-phase diffusions for simulations of Li-ion batteries. J Power Sources 556:232413. https://doi.org/10.1016/j.jpowsour.2022.232413
79. Hutzenlaub T, Thiele S, Paust N, Spotnitz R, Zengerle R, Walchshofer C (Jan. 2014) Three-dimensional electrochemical Li-ion battery modelling featuring a focused ion-beam/scanning electron microscopy based three-phase reconstruction of a LiCoO2 cathode. Electrochim Acta 115:131–139. https://doi.org/10.1016/J.ELECTACTA.2013.10.103
80. Kim G-H, Pesaran A, Spotnitz R (2007) A three-dimensional thermal abuse model for lithium-ion cells. J Power Sources 170(2):476–489. https://doi.org/10.1016/j.jpowsour.2007.04.018
81. Hatchard TD, MacNeil DD, Basu A, Dahn JR (2001) Thermal model of cylindrical and prismatic lithium-ion cells. J Electrochem Soc 148(7):A755. https://doi.org/10.1149/1.1377592
82. Golubkov AW et al (2015) Thermal runaway of commercial 18650 Li-ion batteries with LFP and NCA cathodes – impact of state of charge and overcharge. RSC Adv 5(70):57171–57186. https://doi.org/10.1039/C5RA05897J
83. Ouyang D, Chen M, Wang J (2019) Fire behavior of lithium-ion battery with different states of charge induced by high incident heat fluxes. J Therm Anal Calorim 136:2281–2294. https://doi.org/10.1007/s10973-018-7899-y
84. Liu X et al (2018) Thermal runaway of lithium-ion batteries without internal short circuit. Joule 2(10):2047–2064. https://doi.org/10.1016/j.joule.2018.06.015
85. Wang Z, Ouyang D, Chen M, Wang X, Zhang Z, Wang J (2019) Fire behavior of lithium-ion battery with different states of charge induced by high incident heat fluxes. J Therm Anal Calorim 136:2239. https://doi.org/10.1007/s10973-018-7899-y
86. Kim J, Mallarapu A, Finegan DP, Santhanagopalan S (2021) Modeling cell venting and gas-phase reactions in 18650 lithium ion batteries during thermal runaway. J Power Sources 489:229496

87. Energy Storage Integration Council (ESIC) Energy Storage Reference Fire Hazard Mitigation Analysis. EPRI, Palo Alto, CA: 2019. 3002017136

88. Walker W, Finegan DP, Shearing PR, Battery failure databank (Last accessed October 2022). https://www.nrel.gov/transportation/battery-failure.html

89. Finegan DP et al (2021) The application of data-driven methods and physics-based learning for improving battery safety. Joule 5(2):316–329. https://doi.org/10.1016/J.JOULE.2020.11.018

90. Feng X, Pan Y, He X, Wang L, Ouyang M (Aug. 2018) Detecting the internal short circuit in large-format lithium-ion battery using model-based fault-diagnosis algorithm. J Energy Storage 18:26–39. https://doi.org/10.1016/J.EST.2018.04.020

91. Hu G, Huang P, Bai Z, Wang Q, Qi K (Nov. 2021) Comprehensively analysis the failure evolution and safety evaluation of automotive lithium ion battery. eTransportation 10:100140. https://doi.org/10.1016/J.ETRAN.2021.100140

92. Barnett B, Ofer D, Sriramulu S, Stringfellow R (2012) Lithium-ion batteries – safety. In: Encyclopedia of Sustainability Science and Technology, Meyers RA (ed). Springer, New York

93. Srinivasan R, Demirev PA, Carkhuff BG, Santhanagopalan S, Jeevarajan JA, Barrera TP (2020) Review—thermal safety management in Li-ion batteries: current issues and perspectives. J Electrochem Soc 167(14):140516. https://doi.org/10.1149/1945-7111/ABC0A5

94. Kim J, Mallarapu A, Yang C, Santhanagopalan S (2021) Modeling cell venting and gas-phase reactions in lithium-ion cells during thermal runaway. Presented at the AABC 2021.

95. Torres-Castro L, Kurzawski A, Hewson J, Lamb J (2020) Passive mitigation of cascading propagation in multi-cell lithium ion batteries. J Electrochem Soc 167(9):090515. https://doi.org/10.1149/1945-7111/ab84fa

96. Kamyab N, Coman PT, Reddy SKM, Santhanagopalan S, White RE (2020) Mathematical model for Li-s cell with shuttling-induced capacity loss approximation. J Electrochem Soc 167(13):090534. https://doi.org/10.1149/1945-7111/abbbbf

97. Sethuraman VA, Srinivasan V, Bower AF, Guduru PR (2010) In situ measurements of stress-potential coupling in lithiated silicon. J Electrochem Soc 157(11):A1253. https://doi.org/10.1149/1.3489378

98. Sethuraman VA, Hardwick LJ, Srinivasan V, Kostecki R (2010) Surface structural disordering in graphite upon lithium intercalation/deintercalation. J Power Sources 195(11):3655–3660. https://doi.org/10.1016/J.JPOWSOUR.2009.12.034

99. Cheng X-B, Zhang R, Zhao C-Z, Zhang Q (2017) Toward safe lithium metal anode in rechargeable batteries: a review. Chem Rev 117(15):10403–10473. https://doi.org/10.1021/ACS.CHEMREV.7B00115

100. Chen Y et al (2021) A review of lithium-ion battery safety concerns: the issues, strategies, and testing standards. J Energy Chem 59:83–99. https://doi.org/10.1016/J.JECHEM.2020.10.017

Chapter 7
Accelerating Battery Simulations by Using High Performance Computing and Opportunities with Machine Learning

Srikanth Allu, Jean-Luc Fattebert, Hsin Wang, Srdjan Simunovic, Sreekanth Pannala, and John Turner

Abstract The US Department of Energy (DOE) estimates that the battery pack cost of $60 per kilowatt-hour while increasing the driving range to over 300 miles and vehicle charging under 15 min or less would enable mass penetration of electric vehicles in the USA by 2030. These projections are based on the currently available high-density cell chemistry combined with a system level design and optimized electrodes. Electrodes for current state-of-the-art lithium-ion cell technology are fabricated from electrode slurry comprising of active material, polymeric binder, and conductive diluent such as carbon black that are coated on metal current collectors such as copper and aluminum. Given these challenging requirements for development of electrical energy storage devices for future transportation needs, a predictive simulation capability which can accelerate design by considering performance and safety implications of different geometry, materials, and chemistry choices is required. In this chapter, we discuss our state-of-the-art three-dimensional modeling framework, providing examples on battery performance and safety simulations. We also present approaches to use machine learning in the context of large datasets that are becoming available.

S. Allu (✉) · J.-L. Fattebert · H. Wang · S. Simunovic · J. Turner
Computational Sciences and Engineering Division, Oak Ridge National Laboratory, Oak Ridge, TN, USA
e-mail: allus@ornl.gov

S. Pannala
Corporate Technology and Innovation, SABIC, Sugar Land, TX, USA

This is a U.S. government work and not under copyright protection in the U.S.; foreign copyright protection may apply 2023
S. Santhanagopalan (ed.), *Computer Aided Engineering of Batteries*, Modern Aspects of Electrochemistry 62, https://doi.org/10.1007/978-3-031-17607-4_7

7.1 Introduction

Lithium-ion batteries are complex electrochemical systems whose performance and safety are governed by coupled nonlinear electrochemical-electrical-thermal-mechanical processes over a range of space and time scales. Understanding of the role of these processes as well as the development of predictive capabilities for design of better performing batteries requires developing a framework that can integrate the models representing various physical processes and be used as a unified engineering platform to evaluate the battery designs. Under the computer aided engineering for batteries (CAEBAT) program, models at the macroscopic or system level were typically based on electrical circuit models or simple 1D models, and there were no design tools for batteries that could leverage the significant advances in modeling development across DOE, industry, and academia. To this end we have developed a three-dimensional electrochemical-electrical-thermal framework that allows to create 3D cell and battery pack models that explicitly simulate all the battery components (current collectors, electrodes, and separator). The models are used to predict battery performance under normal operations and to study thermal and mechanical safety aspects under adverse conditions. In the subsequent sections we provide an overview of the development of this framework and present studies that have been conducted at various scales for performance as well as safety. The framework can utilize the latest high performance computing technologies which have been used for accelerating the development of new battery designs. In the last Sect. 7.6, we describe some recent developments in the Machine Learning (ML) for modeling physics and chemistry processes of lithium-ion batteries, at various scales that are becoming an important component of analysis for Lithium-Ion batteries.

7.2 Continuum Electrochemical and Thermal Model for Batteries

The equations describing the physical processes are based on conservation of charge and species using Butler–Volmer reaction kinetics [14, 22]. The lithium concentration and electric potentials solutions are calculated for multiple phases by solving conservation equations for species and charge which are constructed using porous electrode and concentrated solution theory. The continuum model is derived by volume averaging approaches that can be extended to full cell with arbitrary 3D electrode structures. In the volume averaging approach [4, 31, 45], the electrode and the electrolyte are considered as superimposed continua over unit volume 1. This assumes that the microscopic features of the electrode be more or less uniform (i.e., can be described by the first moment of their feature distribution) and that their characteristic length scales are much smaller than the size of the representative volume. In this procedure, the effect of microscale structures can be captured by considering anisotropy in the electrode material effective properties

Fig. 7.1 Schematic of a
representative volume

such as diffusivity and conductivity. In addition, this approach can account for
both spatial and temporal variations of properties that are critical to model the
heterogeneity of the electrode structure over its life-time and also effectively model
changes in constituent materials to understand safety mitigation strategies.[1]

Let L be the characteristic length of the averaged volume that is larger than
the secondary particles, i.e., $L >> r$ and normalized weighting function $g(r)$
which is monotonically decreasing over the distance r. The function must satify
the constraint:

$$4\pi \int_0^\infty g(r)r^2 dr = 1 \tag{7.1}$$

Given this function, we can define local spatial averages of interest to construct
the volume averaged equations. Consider a representative volume element $d\Omega$ over
which the quantities of interest are averaged as shown in Fig. 7.1. This allows one
to define,

$$\langle \phi \rangle = \int_\Omega \phi g(r) d\Omega \tag{7.2}$$

Next, we derive the individual conservation equations for the electrodes and
electrolyte averaged over a volume larger than the individual particles but smaller
than the electrode dimensions or the diameter of the secondary particles—as the
case might be for definition of the device scale. If dV is the representative volume
containing both the solid and electrolyte phases,

then, omitting any convective flow within the electrolyte, these equations can be
rewritten as follows:

$$\left\langle \frac{\partial \tilde{\psi}}{\partial t} \right\rangle = \epsilon \frac{\partial \psi}{\partial t} \tag{7.3}$$

$$\langle \nabla \tilde{\psi} \rangle = \epsilon \nabla \psi \tag{7.4}$$

---

[1] Significant portion of the derivations presented in this section were first published in [2] and are
reproduced with permissions from Elsevier.

Species Concentration in the Electrolyte Concentrated solution theory is used to model transport of ions in the electrolyte as the presence of individual ions influences the net electric fields. Species conservation for lithium-ions (molar concentration c_e) within the electrolyte is given by:

$$\frac{\partial(\epsilon_e c_e)}{\partial t} - \nabla \cdot \left(\epsilon_e D_e^{eff}(\epsilon_e) \nabla c_e \right) = \frac{1 - t_+^0}{F} j^{Li} \tag{7.5}$$

where the diffusion coefficient D_e is a function of the salt concentration. The effective diffusion coefficient is computed using the relation $D_e^{eff}(\epsilon_e) = D_e \epsilon_e^{0.5}$. The right hand term of the equation represents the rate at which the ions are entering or leaving the electrolyte due to electrochemical reactions. The transference number t_+^0 is a measure of the intercalation efficiency [15].

Species Concentration in the Electrode Transport of Li-ions in the active material is the rate limiting process. This phenomenon occurs at a length scale that is different than the characteristic length for the remainder of the problem. The current prevailing research approach is to assume that the particles are uniform spheres. Correspondingly, the solid phase diffusion equation is solved in spherical coordinates at each volume element across the thickness of the electrode. Approximations of analytical solutions from Duhamel superposition method based on diffusion lengths are available. These relations are useful for upscaling from the particle scale to evaluate solid phase surface concentrations at higher length/timescales. Using the volume averaging techniques proposed above we write the mass balance for solid phase,

$$\frac{\partial(\epsilon_s c_{s,avg})}{\partial t} - \nabla \cdot \left(\epsilon_s D_s^{eff}(\epsilon_s) \nabla c_{s,avg} \right) = -\frac{j^{Li}}{F} \tag{7.6}$$

where $D_s^{eff}(\epsilon_s) = D_s \epsilon_s^{0.5}$ and the closure relation given by,

$$c_s = c_{s,avg} + \frac{j^{Li} R_s}{5 a_s F D_s} \tag{7.7}$$

Conservation of Charge in Electrolyte Phase Under the assumption of electroneutrality and absence of large electric fields in electrolyte phase, accumulation of charge and discharge at the electric double layer is neglected and the conservation of charge is instantaneous. Based on volume averaging principles from [14, 41] the conservation of charge along with Ohm's law that relates the gradient of electric potential to the electronic current density in the electrolyte phase gives the following equation.

$$\nabla \cdot (\epsilon_e \kappa^{eff}(\epsilon_e) \nabla(\phi_e)) + \nabla \cdot \left(\epsilon_e \kappa_D^{eff}(\epsilon_e) \nabla ln c_e \right) = -j^{Li} \tag{7.8}$$

where $\kappa^{eff}(\epsilon_e)$ and $\kappa_D^{eff}(\epsilon_e)$ are ionic and diffusive conductivity given by,

$$\kappa^{eff}(\epsilon_e) = \kappa\epsilon_e^{0.5} \tag{7.9}$$

$$\kappa_D^{eff}(\epsilon_e) = \frac{2RT\kappa\epsilon_e^{0.5}}{F}(t_+^0 - 1)\left(1 + \frac{dlnf_+}{dlnc_e}\right) \tag{7.10}$$

The transference number t_+^0 is defined as the fraction of the current that is carried by ion in an electrolyte solution of uniform composition.

Conservation of Change in Electrode Phase The volume average equation for conservation of charge using Ohm's law in solid phase is given by

$$\nabla \cdot (\epsilon_s \sigma^{eff}(\epsilon_s)\nabla\phi_s) = j^{Li} \tag{7.11}$$

where $\sigma^{eff}(\epsilon_s) = \sigma$ is the effective solid phase conductivity. The first term represents the charge transport due to electronic conduction.

Chemical Kinetics The Butler–Volmer kinetics equation gives the total transfer current at the electrode/electrolyte interface which is the difference between the rate of forward reaction to the rate of backward reaction. The α_{cj} and α_{bj} correspond to the fraction of the applied potential that promotes cathodic and anodic reactions.

$$i_{nj} = i_0 \left[e^{(\frac{\alpha_{aj}F}{RT}\eta_j)} - e^{-(\frac{\alpha_{bj}F}{RT}\eta_j)} \right] \tag{7.12}$$

The potential at which the net rate is zero, i.e., rate of forward reaction is equal to the rate of backward reaction is known as the equilibrium potential, and the difference between the actual potential and equilibrium potential is given by the surface overpotential,

$$\eta_j = \phi_s - \phi_e - U_{j,ref} - i_{nj}R_f \tag{7.13}$$

The exchange current density is defined as a function of rate constants for anodic and cathodic reactions and species concentrations.

$$i_0 = kc_e^{\alpha_{aj}}(c_{s,max} - c_s)^{\alpha_{aj}}c_s^{\alpha_{cj}} \tag{7.14}$$

$$j^{Li} = \begin{cases} a_{s1}i_{n1} & \text{Cathode} \\ 0 & \text{Separator} \\ a_{s2}i_{n2} & \text{Anode} \end{cases} \tag{7.15}$$

State of charge is given by $\theta = c_s/c_{s,max}$

The energy balance equation is the three-dimensional heat conduction equation:

$$\rho C_p \frac{\partial T}{\partial t} - \Delta(k\Delta T) = q \tag{7.16}$$

The anisotropy of thermal properties is due to the repeating layers in the cell structure. The thermal conductivity in-plane is several times higher than that through-plane. The energy balance Eq. 7.16 is solved across the cell geometry, and the heat source, q, varies with the domain. In the interconnects and current collector foils the heat source is a simple Ohmic heating model. Detailed heat generation terms within the dual-electrode insertion electrodes were derived by Bernardi et al.67 who discuss additional heating terms within the cell. In its simplified form, which is commonly employed in battery simulations, the heat generation consists of irreversible energy losses, reversible entropy change due to particular half-cell reactions and ohmic heating.

$$q = \sum a_j i_j (\phi_s - \phi_e - U) + \sum a_j i_j T \frac{\partial U_j}{\partial T} + \frac{i_s^2}{\sigma_{eff}} \tag{7.17}$$

The summation over all reactions $j = 1 \ldots M$ in the case of Li-ion intercalation systems simplifies to two half-cell reactions. The OCP vs. lithium are denoted as U_j.

Computational Methodology and Implementation The initial value problem for this system of differential-algebraic equations (DAEs) with dependent variables $\psi = \{c_s, c_e, \phi_s, \phi_e, j_{Li}\}$ can be written as:

$$F(t, \dot{\psi}, \psi) = 0 \tag{7.18}$$

The initial conditions are $\psi(0) = \psi_0$ and $\dot{\psi}(0) = \dot{\psi}_0$. In this system of equations the charge conservation equations are algebraic constraints making it an index 1 DAE system. Numerically time integrating this system requires a consistent initial solution for the algebraic constraints at time t_0. To compute the solution of these DAEs, the pair of initial vectors ψ_0 and $\dot{\psi}_0$ that satisfies the Eq. (7.18) is an essential prerequisite. Next, we describe techniques used to initialize and the solvers for adaptive time integration.

The dependent variables in this formulation are $\psi = \{c_s, c_e, \phi_s, \phi_e, j_{Li}, T\}$. We can use same order of approximation for all dependent variables and the finite element approximation can be written as,

$$\psi^h = [N(x, y, z)] \psi \tag{7.19}$$

where ψ^h is the finite element approximation of the dependent variables and ψ is the vector of nodal degrees of the freedom.

It has been well established that the problem of inconsistent initial conditions do cause numerical solution techniques to fail for an index one system of differential-algebraic equations (DAEs). Inconsistent initializations are caused in a drive cycle due to alternating boundary conditions depending on the charging and discharging of a battery. Under a galvanostatic discharge a constant flux condition and a reference potential are imposed on the non-adjacent boundaries. At any instant t, the assumed constant initial solution profiles are inconsistent with the boundary conditions. It is a non-trivial task to consistently set the initial solution by trial and error method. To avoid this issue, the system of equations is recast into a system of steady state equations by change of variables with $\dot{c} = y$ and $c(x, y, z) = c^{const}$.

$$0 = f(0, y, c^{const}, \phi, j_{Li}, T) \tag{7.20}$$

$$0 = g(0, c^{const}, \phi, j_{Li}, T) \tag{7.21}$$

where c, ϕ, j_{Li}, T are vectors of differentiable and algebraic variables. The solutions to these equations will be consistent with the boundary conditions. This initialization computation is achieved by solving the nonlinear system using a Newton–Krylov solution technique. This DAE re-initialization technique should be used whenever there is a sudden jump in the current boundary condition. These variable current profiles are very useful during the pulse-testing or drive-cycle testing of the battery.

For the time integration scheme used is the variable order, variable coefficient BDF in the fixed leading coefficient form that is implemented in the IDA module of the Suite of Nonlinear and Differential/Algebraic Equation Solvers (SUNDIALS) [24]. The order ranges from 1 to 5 of the BDF given by the formula,

$$\sum_{i=0}^{q} \alpha_{n,i} \psi_{n-i} = h_n \dot{\psi}_n \tag{7.22}$$

where ψ_n and $\dot{\psi}_n$ are the computed approximations to $\psi(t_n)$ and $\dot{\psi}(t_n)$, respectively, and the step size is given by $h_n = t_n - t_{n-1}$. The coefficients $\alpha_{n,i}$ are uniquely determined by the order q, and the recent history of the step sizes. Evaluating equation (7.18) at $t = t_{n+1}$ and using (7.22) to replace $\dot{y}$ leads to a nonlinear algebraic system to be solved at each step:

$$F\left(t_n, \psi_n, h^{-1} \sum_{i=0}^{q} \alpha_{n,i} \psi_{n-i}\right) = 0 \tag{7.23}$$

Newton iteration applied to the nonlinear DAEs generates a sequence of approximations ψ_k to ψ^*, where $\psi_{k+1} = \psi_k + s_k$ and the Newton step s_k is the solution to the system of linear equations

$$F'(\psi_k)s_k = -F(\psi_k) \tag{7.24}$$

where F' is the Jacobian of F evaluated at ψ_k. The Jacobian J is some approximation to the system

$$J = \frac{\partial F}{\partial \psi} + \alpha \frac{\partial F}{\partial \dot{\psi}}$$ (7.25)

where $\alpha = \alpha_{n,0} / h_n$.

We use Inexact Newton iteration with scaled preconditioned Generalized Minimum Residual Method (GMRES) as the linear solver using the matrix-free product of current Jacobian (Jv). In this Generalized Minimum Residual Method (GMRES) [38] solution to linear systems is obtained by projecting onto the Krylov subspace

$$K(A, v) = span\{v, Av, A^2v, \ldots, A^{m-1}v\}$$ (7.26)

where $v = f(x_{n+1})$. Arnoldi's method [10], which requires only the action of A on a sequence of vectors, is used to construct an orthonormal basis for this space.

The Jacobian J defined in 7.24 is computed by IDA internally by difference quotients approximated by

$$J = \frac{[F(t, \psi + \sigma v, \dot{\psi} + \alpha \sigma v) - F(t, \psi, \dot{\psi})]}{\sigma}$$ (7.27)

where the increment $\sigma = 1 / ||v||$

Among the various Krylov methods available, GMRES is selected because it guarantees convergence with nonsymmetric, nonpositive definite systems [38] although it is expensive and takes more iterations per Jacobian solve. In order to reduce these GMRES iterations, we improve the condition number of the Jacobian matrix by preconditioning the problem. Left preconditioning is utilized resulting in following system

$$P^{-1}J\delta y = -P^{-1}F.$$ (7.28)

For preconditioner solve we use Algebraic multigrid (Trilinos ML) V-cycle solver with a coarse grid solver [23].

7.3 Novel Battery Architectures

At the fundamental level, highly complex assemblies of scalable three-dimensional electrode and device architectures with good rate capabilities, life, and safety are governed by hierarchical multi-scale interactions between materials, electrochemical transport (diffusion and kinetics), electrode architectures, and system design. Significant increase of battery energy density needs to embrace three important metrics: performance, cost, and safety. Under such design constraints a single layer electrode is limited by the both energy and power due to electronic and ionic transport processes [51]. One of the key prerequisites for success of future lithium-ion technology is an increase of the energy density per unit mass or volume at the cell or system level without sacrificing on the power performance. One of the several approaches would be to design thicker electrodes without sacrificing the gravimetric energy density and maintaining the power performance through efficient electronic and ionic pathways [6]. Another radical approach would be to move towards a 3D based architecture to pack more energy density into limited space and reduce the transport distances to maintain the high power density [28]. 3D architectures are electrode and battery configurations that are non-planar configurations and offer the potential for greatly improved performance, particularly in terms of volumetric energy density, compared to traditional planar configurations [28, 36]. In this work, we present a computational study 3D electrode and device architecture and demonstrate enhancements in energy density even using the current industry standard cathode materials such as $LiCoO_2$ or $LiFePO_4$ or NMC. Further, this could be improved by moving to Li-metal anode, having higher tap density of conventional cathode materials and lower porosity sustained by 3D architectures without loss of transport properties that effect power density. These designs are expected to be scalable and to operate over length scales that exhibit rapid electron and ion transport.

Current State of The Art and Need for 3D Electrode Structures: Conventional Li-ion planar electrodes have a host of material, transport, and design limitations that prevents concurrent high energy and power density. The porosity, the thickness, and the homogeneity of the electrode are only partly controlled by manipulating the slurry composition, viscosity, casting, and drying parameters. Overall this results in the electrode materials randomly distributed along the electrode thickness surrounded by uneven distribution of carbon and the polymeric binder. From an electrochemical transport perspective this gives rise to a number of issues such as (i) poor electrolyte access to redox-active materials, (ii) concentration polarization at high discharge/charge rates, and (iii) anisotropic thermal transport and mechanical stress–strain effect during intercalation. In addition the electrode fabrication process intrinsically has other micron scale inhomogeneities combined with presence of passive components and additives that has penalty in terms of energy density. In this study we report the interdigitated electrode and cell architecture that have high areal and power densities that in principle can be scalable.

7.3.1 Application to Non-planar Interdigitated Geometry

The focus so far has been to maximize electrode capacity utilization and rate performance at individual electrode level but not necessarily at the cell level for maximizing the energy density. We perform detailed computational modeling on selected 3D geometries to maximize the energy density per unit area (and volume) ensuring shorter electron and ion paths for rapid kinetics and minimum

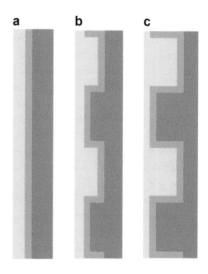

Fig. 7.2 Three cell sandwich geometries compared in simulations: (**a**) planar; (**b**) shallow trench; (**c**) deep trench (depth is 2x of shallow trench). Adapted from [2] with permission from Elsevier

electrode/electrolyte interfacial resistance. The model 3D electrode structure can be in principle repeated to build thicker electrodes without loss of electrochemical performance and design features.

To evaluate 3D battery architectures, we maintain the electrode volume constant, resulting in comparable specific energy densities. This is accomplished by maintaining the height and width of the cell constant while varying the electrode thickness. We reduce the electrode thicknesses in the planar configuration and introduce trenches of equal height (to accommodate a separator with constant thickness) on either electrode. See Fig. 7.2. The trench depth is doubled between the shallow (Fig. 7.2b) and deep configurations (Fig. 7.2c). The energy density is constant across the three configurations; the surface area exposed to initial reaction fronts increases. Since the energy density across the three cases is the same, the discharge current for the boundary conditions at various C-rates remains the same.

The material balance between the positive and negative electrodes should be maintained as well, i.e., the number of trenches on the either side of the cell sandwich should be the same. In Fig. 7.3 we show the comparison of the solid phase concentration profiles among three configurations at the end of 1C discharge. The skewness in the concentration distribution is the result of un-symmetric design discussed above. Lithium content in the solid phase at the end of 5C discharge is shown in Fig. 7.3. The active material is utilized primarily in the region near to the separator in planar configuration (Fig. 7.3a) with significant concentration gradient developed across the positive electrode. Better material utilization (lithium concentration in solid electrodes) can be seen in interdigitated designs in Fig. 7.3b, c.

The interdigitated designs of cell sandwich show higher discharge capacity due to better utilization of the electrode material. The cell capacity increases with the depth of the trenches. However, the specific energy density remains the same. Traditionally, Ragone plots are used to evaluate the power/energy characteristics

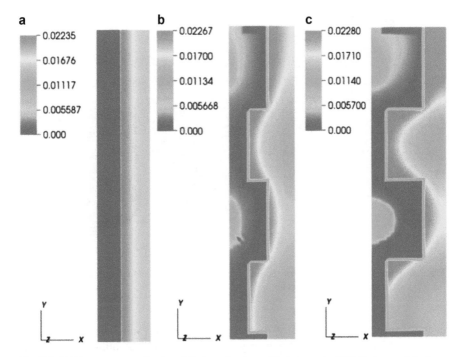

Fig. 7.3 Lithium concentration in solid phase (mol mm-3) at the end of 5C discharge: (**a**) planar geometry; (**b**) shallow trench; (**c**) deep trench (depth is 2x of shallow trench).Adapted from [2] with permission from Elsevier

of an energy storage device. These plots compare volumetric energy density against power. As discussed above, the interdigitated design increases usable capacity from the cell. Due to better electrode material utilization, the power density of the cell increased with depth of the trenches, confirming that 3D architectures for cells with higher power characteristics is promising.

7.4   Thermal Management

For safe and reliable battery operation, regulation of battery temperature is of the paramount importance. Due to the limitations of the thickness of the coatings and battery form factors, various modular designs of these battery packs are required. This modular design is heavily dictated by the advancement of the temperature sensing and thermal management systems. The primary challenge is to understand the heat release under variable drive cycle and the ability to dissipate heat and maintain uniform temperature at all times. Non-uniform temperature distribution across the cell and module has implications on the under-utilization of active materials can accelerate the capacity degradation. This leads to localized aging and

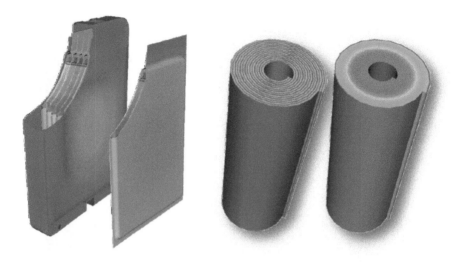

Fig. 7.4 Temperature distribution of the full scale pouch modules and cylindrical cells. Adapted from [1] with permission from Elsevier

eventually reduced cycle life of the cell. So far, all the simulation studies have always used lumped models for heat generation to study the influence of liquid-based or convective cooling at the surfaces. But because of the thermal gradients within the cell, it is necessary to study the electrochemical response coupled to the spatial temperature distribution considering all the electrode components. Based on the formulation presented in the previous chapter, we have developed an efficient scalable parallel simulation framework for solving coupled multi-domain, electrochemical and thermal transport in batteries (Fig. 7.4). The interaction between the adjacent components that are in contact can be modeled using the mixed boundary conditions. The contact surface roughness can also be modeled with in the current framework.

Large EV packs usually employ active cooling systems, such as liquid cooling or forced air as a thermal management strategy. They are feasible for large EV packs where space is not critical. For cost effective and compact battery designs, a system approach from the cell level to the pack level is needed. At the cell level, an ideal method for thermal management would be to regulate the internal regions of the cell directly at the heat source. Dissipation of heat through metal current collectors can be used for cooling as an alternative to the standard practices of using cooling plates. Battery performance and life span can be improved by maintaining a constant internal cell temperature. Integrating thermal regulation at the cell level, we could reduce the system level overhead needed to achieve higher power densities. The control system overhead can also be further reduced because the thin profile of current pouch cells –a form factor dictated by thermal diffusion rates– could be optimized around a new set of parameters beyond existing thermal limits of the cells.

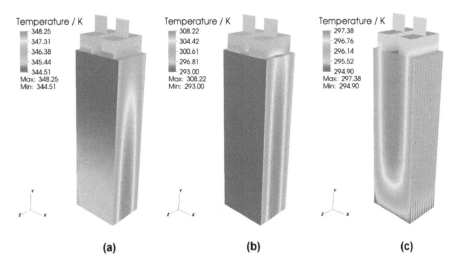

Fig. 7.5 Temperature distribution in a thick prismatic cell with (**a**) no cooling applied; (**b**) cooling of the pouch surface; (**c**) cooling of the current collectors. Adapted from [32] with permission from AIP Publishing LLC

In order to demonstrate this concept, a thick cell containing 84 cell sandwiches across the thickness was modeled. The electrochemical model parameters and discharge currents that correspondingly to produce a 5C discharge rate can be found in reference [32]. Figure 7.5 shows the comparison of the coupled electrochemical and thermal solution under various cooling boundary conditions. Idealized liquid cooling was simulated by holding the corresponding cell surfaces at 295 K (Dirichlet boundary conditions). The rest of the surfaces had convective air cooling modeled by a convective heat transfer coefficient equal to $15 \, \text{W/m}^2 \text{K}$. The cell with the new design allowing core cooling via current collectors is effective at controlling the temperature range and uniformity (Fig. 7.5c). The benefits of cell liquid cooling using the foils are quantified using the through-thickness temperature profiles in Fig. 7.6 that shows the same cooling scenarios as in Fig. 7.5. Applying liquid cooling to the current collectors results in the lowest temperature across the cell (Fig. 7.6c). Using extended metal foils for cooling the battery from sides not only facilitates manufacturing of thicker cells but also reduces the thermal gradients across the cell thickness by direct heat removal from the core of the battery.

7.5 Safety

One of the prime concerns with increasing number of electrical vehicles on the road is the battery response to the mechanically induced internal short circuit. Due to the lack of physical evidence after thermal runaway events, the final failure

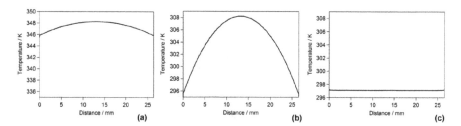

Fig. 7.6 Temperature distribution in a thick prismatic cell with (**a**) no cooling applied; (**b**) cooling of the pouch surface; (**c**) cooling of the current collectors. Adapted from [32] with permission from AIP Publishing LLC

mechanism(s) are hard to pinpoint. Internal short circuit induced by the mechanical deformation is believed to be the main cause of battery failure and thermal runaway in crashes involving electric vehicles. In order to study this, mechanical indentation tests at the cell level and stack level have been conducted. Abuse tests such as cell crushing or pinching help evaluate the final outcome of the failure conditions. However, these tests often result in a thermal runaway with not much evidence to examine. To understand mechanical failure of materials within the cells, we performed incremental indentations on the prismatic cells in aluminum cans. The indentation tests were stopped at various stages and the hard case aluminum cans were able to retain the deformation after the tests. Post-mortem analysis using cross-section imaging and X-ray tomography was performed. The folding of the different layers before failure and the indentation depth at failure were identified.

Copper current collectors fragmented after spherical indentation on both commercial prismatic as well as large format pouch cells, as observed by 3D X-ray computed tomography (XCT). Microscopic analysis using scanning electron microscopy (SEM), scanning transmission electron microscopy (STEM), and X-ray photoelectron microscopy (XPS) on copper current collectors from used commercial cells and pristine anodes showed rough interface areas affected by reactions and diffusion in the used cell. Electron probe micro-analysis (EPMA) element mapping attributed such deterioration at the interface area to enrichment in oxygen and phosphorus content. Phosphorus was also distributed uniformly inside the current collector. These distributions were confirmed in STEM analysis. XPS depth profiles revealed the presence of Li, F, P, O, and C and diffused at least 50 nm into the copper resulting in embrittlement of the foils, compared to the pristine anode which showed a very smooth C/Cu interface.

The indented prismatic and large format cells were examined by 3D XCT after a 0.1 V drop in the open circuit voltage signaling an initial short circuit. Figure 7.7a–c are images from three different orientations. The cobalt in the cathodes shows up as thicker layers with a light grey color. Since the carbon anodes are X-ray transparent, only the copper foils are visible. As noted earlier [14], the copper layers were fragmented, and they showed as discontinuous lines in the cross-sectional views (Fig. 7.7a,b), but as mud-cracks when viewed in the plane parallel to the layers

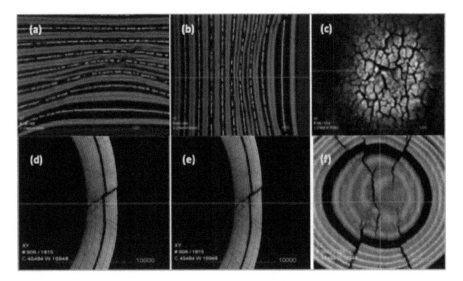

Fig. 7.7 Fragmentation of copper current collectors observed by 3D XCT in a small prismatic cell (**a**), (**b**), and (**c**), and the top cell of large format LG-Chem 10-cell stack (**d**), (**e**), and (**f**) after indentations. Adapted from [46] with permission from Elsevier

(Fig. 7.7c). For the large format cell, the top cell in the 10-cell stack was selected for XCT. An external DC power source was connected to the top cell and voltage drop was monitored from the positive and negative tabs. Because of the restriction of sample holder to scan a 5.5 mm × 50 mm × 145 mm strip, spatial resolution was not very high in cross-sectional images Fig. 7.7d,e. However, the discontinuous copper layers were also observed. In the plane parallel to the layers, Fig. 7.7f, the layers are shown as rings due the curvature from indentation. The copper current collectors had the same mud crack patterns as the prismatic cells. The indentation of 10-cell stack has conditions of realistic battery backing in an electric vehicle battery module. The displacement magnitude was a few times the cell thickness because of the deformable cells underneath. Large cracks through the cell have developed. However, the evidence of fragmentation of copper layers was very clear. Because the aluminum layers in the cathode did not show any sign of fragmentation under spherical indentation, the focus of this study was on the copper current collector.[2]

The main criterion for the safe failure mode is to electrically isolate the electrodes and avoid thermal runaway in the event of the internal short due to induced deformation. During mechanical deformation of the battery various material components can undergo a collective or individual breakage. To understand the influence of the electrode fragments that are in partial contact with the rest of the cell sandwich we have conducted electrochemical simulations with a series of configurations

---

[2] The results presented in this section were first published in [46] and are reproduced with permissions from Elsevier.

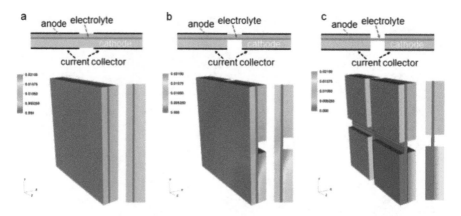

Fig. 7.8 Contour plots of the lithium-ion concentration (mol/cm³) in the solid phase across the segmented electrodes during the end of discharge process. Adapted from [30] with permission from Elsevier

[30]. Under this failure scenario, various possible cell sandwich configurations that we want to investigate are pre-defined and the simulations are setup with appropriate conditions. The various configurations that were constructed are (1) isolated current collectors but electrode materials are intact (2) current collectors and cathode electrodes isolated with anode electrode in place (3) complete isolation of current collectors and electrodes. Assuming that these are the states the battery will end up after deformation, we will apply constant current conditions to the isolated cathode current collector regions and evaluate if the rest of the regions participate in the electrochemical transport. This numerical setup is an idealized version of the internal short where the currents going through the shorted area can vary in exponential form and there could be more dissolution and decomposition reactions that can lead to thermal runaway. This preliminary analysis suffices for the study of electrical isolation and participation in thermo-electrochemical transport. In (Fig. 7.8), we show contour plots of the solid phase Li concentration (mol/cm3) computed across the regions towards the end of discharge under same color scale. For better analysis, all the solutions are plotted with same color scale, i.e., red indicating maximum solid phase concentration attained among all simulations and blue indicating zero value. The cross-section views on the top highlight the isolated current collectors, the separator region filled with electrolyte and the electrodes with colors referring to the initial concentration state. The first case shows that even though the current collectors are completely isolated, the shorted island with constant current boundary condition drives the transport of the entire electrodes. This indicates that if the electrode-electrolyte system is contiguous after deformation, even though we have an isolated current collector islands, it is possible that the short can lead to a thermal runaway as the rest of the electrodes continue to contribute to the energy release. In the second case, where the cathode and the aluminum foils are completely isolated, the entire anode with an isolated copper

foil can still contribute to the transport and subsequent thermal runaway. Finally, we investigate the discharge of a completely isolated anode and cathode. The electrodes segments that are directly across each other contribute to the charge transport. Lithium concentration develops across the separator/cathode interface. But there is no ion flux in adjacent blocks, i.e., the electrically isolated quadrants retain the same initial concentration and do not contribute to the discharge capacity. These case studies allude to better safety mitigation by delaying the heat propagation. But to completely avoid the thermal runaway, in the event of short due to deformation, one must electrically isolate not only the current collectors but also the electrodes to eliminate participation in discharge.

7.6 Opportunities for Data and Machine Learning for Li-Ion Batteries

With the success of large-scale distributed training of deep learning models for natural language processing, image classification and an explosion of data collection, ML/AI methods are being increasingly applied to complex physical phenomenon in various application spaces. Most of the complex physical problems cannot be solved using machine learning techniques from data simply because they violate the principles of being computable within Turing limit, learnable and expressible within Vapnik and Tarski, Godel Limit [8]. If any of the underlying limits have not been established, then it is difficult to scientifically explain the machine learning correlations as this could also predict unphysical results. In this section we discuss some of the early adaptors of ML/AI techniques in battery application across various length and time scales and associated challenges. Also, towards the end we will prelude to opportunities that may be possible where physics-based simulations have been too computationally expensive and ML/AI techniques could help bridge the gaps.

7.6.1 Atomistic Modeling

We have known the laws of quantum mechanics that describe the behavior of atoms and electrons at the microscopic level for almost 100 years. A large effort has been devoted since then in trying to solve these equations for realistic problems, for instance, to understand chemistry at the molecular level. The equations of quantum mechanics being clearly too difficult to solve exactly, various levels of approximations and models have been introduced in an effort to address real application challenges.

Li-ion batteries involve many electrochemical and chemical reactions, and a better understanding of them would help us to design and optimize materials for desired applications. The solid–electrolyte interphase (SEI) which is critical to the stability of the material system and long cycle life of Li-ion batteries (LIB) has been the most studied thin film material in chemistry. Yet there are still many unknown aspects of interface speciation, and its impact on interfacial reduction processes, where computer modeling can help. The Electrolyte Genome Project has accelerated the discovery of new useful electrolytes and generates thousands of electrolyte-surface molecular reactions that give rise to the early SEI formation [35]. This has enabled to rapidly conducting high-throughput first-principles calculations and robustly evaluate crucial properties such as solvation structure, self-diffusion coefficients, conductivity with or without additives, as well as the reduction potentials of the majority as well as minority species. High-throughput capabilities have also enabled fast screening in the search for materials with specific properties.

7.6.1.1   First-Principles Atomistic Models

The idea of First-Principles (FP) atomistic models (also called *ab initio*) is to describe a system based on the knowledge of its atomic constituents only, without using any parameter derived from the experiments. One of the most successful model in this domain is Density Functional Theory (DFT). While still including electrons at the quantum level, it treats the electronic interactions as a mean field, allowing calculations of atomistic systems with a few hundred electrons with the computer resources available to most scientists. Atoms are typically modeled as classical particles (Born-Oppenheimer approximation), and solving the DFT equations provides the forces acting on each atom through the electronic structure, thus allowing molecular dynamics (MD) simulations. These calculations can still be computationally expensive as their complexity grows like the cube of the number of electrons N. $O(N)$ complexity solvers involving various approximations and truncations of small matrix elements have been developed to enable larger simulations [10, 16]. Even with large supercomputers, time-to-solution remains relatively long and limits the length of MD simulations to 10–100 ps in practice. For many applications, this is not sufficient and alternative models are needed.

One approach that helps mitigate the cost of these calculations is to replace part of the system, the solvent, with a continuum solvent model [43]. This is usually a much better approach than simulating a system of molecules in vacuum, in particular, for polar solvents such as EC. In addition, Ward et al. [47] have recently used machine learning to replace theses still expensive calculations with a message passing neural networks (MPNNs) model that requires only the molecular graph as an input. After training this model with over 100,000 molecule calculations, they were able to demonstrate a very good accuracy in predicting the solvation energy of a new molecule. This is an example of how machine learning can help to determine quickly, and at a low computational cost, the physical or chemical properties of a particular molecular system, using a model trained with thousands of computationally expensive calculations.

7.6.1.2 Classical Molecular Dynamics

Classical Molecular Dynamics is a much simpler model of atomistic interactions. It attempts to directly describe the forces between the atoms, modeled as classical particles, with analytical expressions (force-field) that depends on distances between atoms, angles between triplets of atoms, and dihedral angles (angle between two half-planes passing through two sets of three atoms, and having two atoms in common). Interactions can be categorized into different forms such as bonded, Van der Waals, Coulomb. Typically, these interactions can be decomposed into short range components, and a long-range Coulomb potential that can be computed separately. This leads to an $O(N)$ computational complexity for N atoms. Together with the relative simple analytical expressions used to evaluate the forces, this allows for simulations with millions of atoms, or even billions of atoms with high-end HPC resources.

There are, however, severe limitations in applying the classical MD. First, accuracy can be limited since it is often not possible to find an analytical force-field that matches well the actual forces in all the possible configurations a system takes during a simulation. Second, atomic bonds are part of a force-field parameterization. This means that not only the strength of an atomic bond, but the fact that there is an atomic bond between a pair of atoms is encoded in the model. Thus chemical reactions cannot occur spontaneously in a classical MD simulation.

One way to circumvent this limitation is to detect on-the-fly when an atomistic system is in a favorable configuration for a chemical reaction to happen, then make this reaction happen (with a specified probability) by changing the force-field (and thus changing the atomic bonds). Such an approach is implemented in the popular MD software LAMMPS [20] and has been used recently to model SEI growth in Lithium batteries [3]. The added benefit of this approach is that reaction rates can be accelerated by making these reactions happen with higher probabilities than they would otherwise.

An alternative is ReaxFF, a model which allows chemical reactions using bond-length and bond-order, combined with a dynamic distribution of charges. While it allows for bond formation and dissociation, its computational cost is an order of magnitude higher than a standard force-field potential. It was recently used to investigate SEI evolution [50].

7.6.1.3 Machine Learning for Determining Atomic Potentials

When running a long molecular dynamics simulation, a lot of data is generated. More specifically atomic configurations, atomic forces, and system energies for many, many time-steps. Given all that data, one can pose a question: if we have enough pre-computed configurations, can we "learn," for instance, what is the energy of a system of atoms as a function of its atomic configuration? A lot of research was carried out in this direction, starting about fifteen years ago with the work of Behler and Parrinello [7] and is summarized in some recent review articles

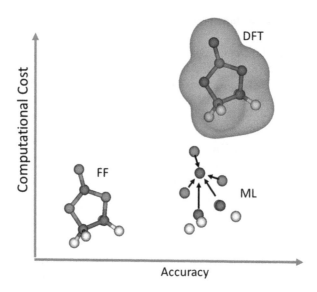

Fig. 7.9 Illustration of various models and their relative accuracy and computational costs. Data-based ML approaches (lower-right) can in many cases provide DFT accuracy at a computational cost close to FF models

[13, 17]. It turns out that in many cases, an atomic potential can be deduced from a set of atomic configurations and their energies. Provided the energy surface obtained through such a supervised machine learning approach is differentiable, one can get the forces acting on all the atoms and generate a molecular dynamics. Neural networks (NNs) and kernel-based regressions are the most common regression models used to construct these potentials. Once the training is complete, evaluating forces with such an ML potential is computationally very cheap compared to a full DFT calculation, with little error compared to an explicit DFT calculation [52] (Fig. 7.9).

Note that unlike ML models that try to directly model the physical or chemical properties of a molecular system, for instance, from its molecular graph, ML atomic potentials enable MD which then can be used to evaluate these properties indirectly. The development of ML potentials is still an evolving field, with enhancement opportunities in the sampling approaches and the atomic environment descriptors. It also has not reached the maturity of DFT or classical FF in the sense that it still requires for a user to train an ML potential for every new application and requires considerable expertise to do that. But the situation is evolving quickly, in particular, with many researchers in the field sharing their data and software with the scientific community.

Some consider the generation of an ML potential from DFT a solved problem, at least for applications where all the interactions are short-ranged. Many applications related to batteries actually do not fall into that category, in particular, due to the long-range Coulomb potential generated by the various anions and cations. But solutions are being developed for charged systems too, for instance, learning not only the energy but also partial charges on each atom which are then used to compute an electrostatic energy to be added to the ML potential [49].

While ML potentials are typically trained using DFT as the underlying model, other even more accurate models could be used in principle. As methodologies evolve, in particular, as better sampling methods are developed that help reduce the number of training configurations, this could become an important benefit from these ML potentials since DFT accuracy is not always sufficient. Note also that the development of ML potentials is harder for simulations that involve liquid electrolytes than it is for solid ones, due to the many configurations, species, and possible interfaces involved [12]. While FP simulations can gives us insight into many phenomena at the atomistic scale, computation involving realistic interfaces such as the anode surface can be quite challenging [29].

7.6.1.4 Chemical Reaction Mechanisms

While FP simulations can in principles be used to model directly chemical reactions, it is not a practical approach due to the long computing time it would take for these reactions to occur spontaneously. Information about individual reactions can, however, be obtained from first principles, in particular, if reactants and products are known. If a reaction path can be guessed, sampling energies and/or forces along that path –using some geometrical constraints, for instance– a thermodynamic integration can be used to compute the free energy difference between reactants and products, as well as the energy barrier for that reaction [11]. Temperature effects can be included by running a finite temperature MD at every sampling point along that path. The bond dissociation energy (BDE) is defined as the enthalpy change for a chemical reaction at 298 K and 1 atm. The reaction kinetics comes from the Arrhenius equation that gives the dependence of the rate constant of a chemical reaction on the temperature T as proportional to

$$e^{-E_a/RT}$$

where E_a is the activation energy and R is the universal gas constant.

As mentioned above, one precondition for these bond dissociation energy and energy barrier calculations is the knowledge of the possibility of that chemical reaction to occur in the first place. For a system such as the electrolyte and its reaction products during the SEI formation, we have some partial knowledge of the possible reactions to expect. But if we consider all the possible reactions, reaction pathways, and products, the number of possibilities grows quickly, and computing all these possible molecules with quantum mechanics methods is prohibitively expensive. Here again ML model can help. Models were proposed for predicting BDEs quite accurately for organic molecules in a fraction of a second after being trained on tens of thousands of small molecules [42, 48]. Similarly, Grambow et al.

[21] developed a deep learning model to quantitatively predict the activation energy given the molecular graphs of a set of reactants and products.

As it is the case for ML atomic potentials (see Sect. 7.6.1.3), a key component of a good ML model is a good descriptor, or vector of input data, that properly encodes molecular information and can be used by an algorithm such as a neural network to learn and predict a feature such as bond dissociation energy or activation energy. A molecular graph can be defined as a set of vertices (atoms) connected by edges (bonds). Following that basic concept, chemically inspired graph neural network (GNN) or Message Passing Neural Networks (MPNNs) [19] were developed and were used quite successfully to predict properties of molecules such as BDEs and activation energies mentioned above, as well as the solvation energies mentioned in Sect. 7.6.1.1.

Leveraging this infrastructure, Samuel Blau et al. [9] have calculated reaction energies and bond dissociation energies in conjunction with automated evaluation of chemical similarities between molecules in the context of SEI formation. Using these reactions, these authors have constructed a chemically consistent graph with stoichiometry constraints that allows for prediction of most probable pathways to desired products in massive reaction networks and study more complex systems than previously possible.

7.6.2 Continuum Scale Electrochemical Systems

7.6.2.1 Mesoscale Structures-Property Relations from Imaging Experiments

The predominant physical phenomena governing the battery at the mesoscale involves electrochemical reactions on active material particle surfaces and ions/electrons transport through the tortuous bi-continuous network formed by the constituent phases. At the electrode scale, machine learning approaches have been utilized to assist the battery imaging data pre-processing, threshold-ing(/segmentation), and quantification and to construct a porous microstructure to conduct numerical simulations and evaluate geometric properties. Existing imaging tools at mesoscale lack the ability to distinguish spatial distributions and anisotropies due to the complex microstructures and various compositional materials of electrodes. Tomography has become a vital tool to characterize and quantify the microstructures and associated properties of porous media, including lithium-ion battery (LIB) electrodes and separators. However, tomographic data often suffer from low image contrast between various material constituents. Pietsch et al. [33] have shown errors induced during data binarization, which translates into uncertainties in the calculated parameters, such as porosity, tortuosity, or specific surface area. Reliable reconstructions of the imaging information obtained of LIBs using X-ray micro tomography are difficult and care must be taken in thresholding and segmentation operations to evaluate specific surface area, porosity,

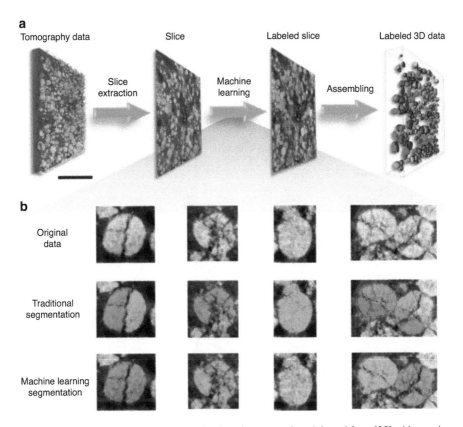

Fig. 7.10 Workflow of the machine learning-based segmentation. Adapted from [25] with permission NatureComm

and tortuosity parameters. Many machine learning methods have been developed that can be trained on these smaller data sets to identify the different material phases and distribution over the data collected from X-ray nano tomography. A small difference in X-ray absorption between epoxy, carbon black, and polymer binder impedes their discrimination. Another emerging method that could be employed are GANs [18] to generate artificial data to supplement real data and reconstruct the missing binder phase which is difficult to detect (Fig. 7.10).

Jiang et al. [25] have developed a machine learning workflow that automatically identifies multiple fragments that broke away from the same particle and quantified the characteristics of every single NMC particles. Using this approach, they have also observed several regions of the NMC particles that are detached from the polymeric binder domain in the severely damaged local regions. Qian et al. [34] developed a data classification method with a nanoresolution synchrotron spectromicroscopy data to investigate asymmetric stress and cause structure disintegration in nickel-rich cathode materials. These authors formulated a machine-learning-

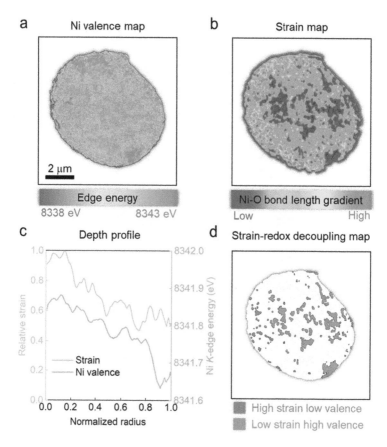

Fig. 7.11 Correlation analysis between Ni valence states and the strain in an NCM 811 particle. (**a**) Ni K-edge energy map from XANES spectroscopy. (**b**) Strain map of the particle from EXAFS spectroscopy processed by machine-learning. (**c**) Correlation between Ni valence and the strain along the depth profile. (**d**) Distribution map of the strain-redox decoupling in the particle. Adapted from [34] with permission from ACS

assisted data clustering approach for reduction of the noise and the dimensionality of the data and further quantified the strain induced lattice distortion (Fig. 7.11).

However, for the data collected at nano scale from synchrotron facilities the collection time is proportional to the sample size, practical samples volumes with representative volume distribution of various materials are inconceivable [39]. Machine learning models trained on these data sets to identify the different material phases could be used to predict the particle phase distribution on the data collected from X-ray micro tomography at higher scale essentially bridging the gap between nano and micron scale. Also, as the data sizes from these science facilities grow the research community would invariably have to deploy these machine learning techniques on large-scale distributed computing that screens and quantifies thousands of particles that supplement the imaging experiment to correlate the structure and properties relations for battery materials.

7.6.2.2 Accelerate the Aging Studies for Cell Level Analysis from Data/Experiments

The deployment of lithium-ion battery at industrial scale needs fast computational models for various dynamic operating conditions. These computer simulations are primarily used to evaluate systems design, active control strategies, and prognosis. The equivalent circuit models which are built on physical characteristics of the charge discharge behavior are used because they provide faster time to solution but are limited in accuracy. The advances in sensor technology and large-scale data collection due to field monitoring enabled new data analytics using machine learning techniques. This has opened the opportunity to predict state of health and cycle life using data-driven methods. Severson [40] has demonstrated a way to extract features from accelerated experiments to identify correlations to capacity degradation for battery lifetime predictions using linear regression model. In this paper, an accelerated aging experimental data is fed as input to a linear model trained via elastic net regression, which extracts features such as incremental capacity change, i.e., $\Delta Q(V)$ from the charging data of the first 100 cycles to predict the final cycle lives of the batteries. Though providing useful information, this ML model is not necessarily field representative and does not encompass all aspects of the dynamic drive cycles and associated degradation modes. Also, Roman Darius, et al. [37] have developed machine learning pipeline by exploring various algorithms that were trained on 179 cells cycled under various conditions with 30 features to calibrate the model. Using the scalable data-driven models, reliable real-time estimation of battery state of health by onboard was presented. In [27] Li, W., et al. observed that recurrent neural nets (RNNs) would be ideal for processing raw time series data but have only a short-term memory of the battery conditions. So, they have developed a supervised learning approach that contains an RNN with long short-term memory (LSTM) cells that can estimate the remaining capacity of a battery under operation. In [26] Weihan Li et al. have attempted to develop physics-based machine learning models by training them with data from physical processes there by eliminating the unphysical and outlier regimes. A validated electrochemical-thermal model was employed to generate a large quantity of data, i.e., voltage, current, temperature, and internal electrochemical states, using a real world operating conditions under various load profiles and temperatures. This data was fed to train an LSTM to estimate the internal concentrations and potentials in the electrodes and the electrolyte at different spatial positions. Even though we have an accurate physical model that can predict accurately but is computationally expensive, this approach of numerical data-based ML model has advantages when you want instantaneous result in real-time. However, from the physical standpoint, the ML based models that are trained on data only that are treated as black box does

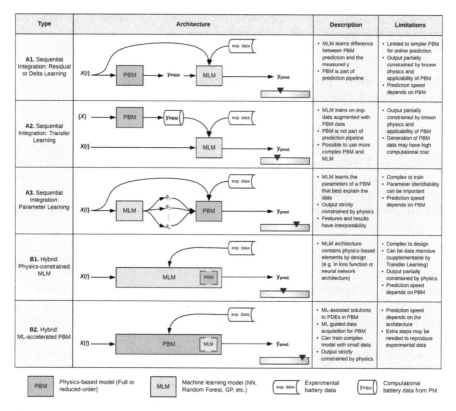

Type	Architecture	Description	Limitations
A1. Sequential Integration: Residual or Delta Learning		• MLM learns difference between PBM prediction and the measured y • PBM is part of prediction pipeline	• Limited to simpler PBM for online prediction • Output partially constrained by known physics and applicability of PBM • Prediction speed depends on PBM
A2. Sequential Integration: Transfer Learning		• MLM trains on exp. data augmented with PBM data • PBM is not part of prediction pipeline • Possible to use more complex PBM and MLM	• Output partially constrained by known physics and applicability of PBM • Generation of PBM data may have high computational cost
A3. Sequential Integration: Parameter Learning		• MLM learns the parameters of a PBM that best explain the data • Output strictly constrained by physics • Features and results have interpretability	• Complex to train • Parameter identifiability can be important • Prediction speed depends on PBM
B1. Hybrid: Physics-constrained MLM		• MLM architecture contains physics-based elements by design (e.g. in loss function or neural network architecture)	• Complex to design • Can be data intensive (supplementable by Transfer Learning) • Output partially constrained by physics • Prediction speed depends on PBM
B2. Hybrid: ML-accelerated PBM		• ML-assisted solutions to PDEs in PBM • ML guided data acquisition for PBM • Can train complex model with small data • Output strictly constrained by physics	• Prediction speed depends on the architecture • Extra steps may be needed to reproduce experimental data

Fig. 7.12 Integration strategies for physics-based models and machine learning models. Adapted from [5] with permission from ECS

not identify or reveal internal physical processes and degradation modes that impact and that could be improved for next generation design.

Recently, an article by Muratahan Aykol et al. [5] described various possible architectures as shown in Fig. 7.12 for integration of physics-based model (PBM) with the ML models which are predominantly used at end of modeling spectrum, respectively. The PBMs are described based on laws of thermodynamics and kinetics which can be used to predict the intricate physical phenomenon for short duration, and on the other hand, the ML models are gaining traction to fast prediction of long duration events such as end of life using data collected from onboard systems.

Recently the authors Hao Tu et al. [44] attempted integration of physical battery phenomena by constructing hybrid models of single particle electrochemical and thermal model with feedforward neural net (FNN) in different configurations. The FNN provides additional information of the internal state of the physical processes and helps to learn more effectively the missing aspects in the measured data.

Many perspective-articles have proposed discussing about combining physics with machine learning, primarily to reduce the dimensionality of data in order to capture all combinations of operating conditions. Even including physics-based models to ensure that we are not violating laws of thermodynamics, in our opinion, it is inevitable to use massive amount of datasets for training machine learning models to encompass all possible field representative use cases. Any of the current models discussed here are not capable of handling Peta bytes of data that are generated on a regular basis in battery testing.

In short span of time, computational science has been increasingly adopted from desktops to a leadership class computing system where simulations can explain and predict electrochemical kinetics and transport phenomena across multiple length and time scales. Over the last decade, computational scientists have developed simulation codes that accelerated battery material discovery using high-throughput screening and electrochemical transport at various scales and have increasingly used a wide array of computing platforms to generate and analyze vast amounts of data. During the same time, experiments from user facilities at several national labs have been increasing the volume of data that is being generated to study the battery materials. The AI/ML technologies are being incorporated in both computational and experimental science efforts. Using these large computing facilities coupled with scalable machine learning and artificial intelligence techniques will greatly accelerate scientific discovery. In particular, coupling AI/ML across the multiple scales can greatly accelerate the material discovery, interfacial stability all the way to device performance of the entire battery system.

Acknowledgments This work is supported by the U.S. Department of Energy's Vehicle Technologies Office at Oak Ridge National Laboratory under Contract No. DE-AC05-00OR22725 with UT-Battelle, LLC. The US Government retains and the publisher, by accepting the article for publication, acknowledges that the US Government retains a non-exclusive, paid-up, irrevocable, world-wide license to publish or reproduce the published form of this manuscript, or allow others to do so, for US Government purposes. The Department of Energy will provide public access to these results of federally sponsored research in accordance with the DOE Public Access Plan (http://energy.gov/downloads/doe-public-access-plan).

References

1. Allu S, Kalnaus S, Elwasif W, Simunovic S, Turner JA, Pannala S (2014) A new open computational framework for highly-resolved coupled three-dimensional multiphysics simulations of Li-ion cells. J. Power Sources 246:876–886
2. Allu S, Kalnaus S, Simunovic S, Nanda J, Turner JA, Pannala S (2016) A three-dimensional meso-macroscopic model for Li-ion intercalation batteries. J Power Sources 325:42–50
3. Alzate-Vargas L, Allu S, Blau SM, ClarkSpotte-Smith EW, Persson KA, Fattebert J-L (2021) Insight into SEI growth in Li-ion batteries using molecular dynamics and accelerated chemical reactions. J Phys Chem C 125(34):18588–18596
4. Anderson TB, Jackson R (1967) Fluid mechanical description of fluidized beds. Equations of motion. Ind Eng Chem Fundam 6(4):527–539

5. Aykol M, Gopal CB, Anapolsky A, Herring PK, van Vlijmen B, Berliner MD, Bazant MZ, Braatz RD, Chueh WC, Storey BD (2021) Perspective—combining physics and machine learning to predict battery lifetime. J Electrochem Soc 168(3):030525

6. Bae C-J, Erdonmez CK, Halloran JW, Chiang Y-M (2013) Design of battery electrodes with dual-scale porosity to minimize tortuosity and maximize performance. Adv Mat 25(9):1254–1258

7. Behler J, Parrinello M (2007) Generalized neural-network representation of high-dimensional potential-energy surfaces. Phys Rev Lett 98:146401

8. Ben-David S, Hrubeš P, Moran S, Shpilka A, Yehudayoff A (2019) Learnability can be undecidable. Nat Mach Intel 1(1):44–48

9. Blau SM, Patel HD, Spotte-Smith EWC, Xie X, Dwaraknath S, Persson KA (2021) A chemically consistent graph architecture for massive reaction networks applied to solid-electrolyte interphase formation. Chem Sci 12(13):4931–4939

10. Bowler DR, Miyazaki T (2012) O(N) methods in electronic structure calculations. Rep Prog Phys 75(3):036503

11. Carter EA, Ciccotti G, Hynes JT, Kapral R (1989) Constrained reaction coordinate dynamics for the simulation of rare events. Chem Phys Lett 156(5):472–477

12. Deringer VL (2020) Modelling and understanding battery materials with machine-learning-driven atomistic simulations. J Phys Energy 2(4):041003

13. Deringer VL, Caro MA, Csányi G (2019) Machine learning interatomic potentials as emerging tools for materials science. Adv Mat 31(46):1902765

14. De Vidts P, White RE (1997) Governing equations for transport in porous electrodes. J Electrochem Soc 144(4):1343–1353

15. Doyle M, Fuller TF, Newman J (1993) Modeling of galvanostatic charge and discharge of the lithium/polymer/insertion cell. J Electrochem Soc 140(6):1526–1533

16. Fattebert J-L, Osei-Kuffuor D, Draeger EW, Ogitsu T, Krauss WD (2016) Modeling dilute solutions using first-principles molecular dynamics: computing more than a million atoms with over a million cores. In: SC '16: proceedings of the international conference for high performance computing, networking, storage and analysis, pp 12–22

17. Friederich P, Häse F, Proppe J, Aspuru-Guzik A (2021) Machine-learned potentials for next-generation matter simulations. Nat Mat 20:750–761

18. Gayon-Lombardo A, Mosser L, Brandon NP, Cooper SJ (2020) Pores for thought: generative adversarial networks for stochastic reconstruction of 3d multi-phase electrode microstructures with periodic boundaries. NPJ Comput Mat 6(1):1–11

19. Gilmer J, Schoenholz SS, Riley PF, Vinyals O, Dahl GE (2017) Neural message passing for quantum chemistry. In: Precup D, Teh YW (eds) Proceedings of the 34th international conference on machine learning, vol 70. Proceedings of machine learning research, pp 1263–1272. PMLR

20. Gissinger JR, Jensen BD, Wise KE (2017) Modeling chemical reactions in classical molecular dynamics simulations. Polymer 128:211–217

21. Grambow CA, Pattanaik L, Green WH (2020) Deep learning of activation energies. J Phys Chem Lett 11(8):2992–2997

22. Gu WB, Wang CY, Li SM, Geng MM, Liaw BY (1999) Modeling discharge and charge characteristics of nickel–metal hydride batteries. Electro Acta 44(25):4525–4541

23. Heroux MA, Bartlett RA, Howle VE, Hoekstra RE, Hu JJ, Kolda TG, Lehoucq RB, Long KR, Pawlowski RP, Phipps ET, et al. (2005) An overview of the Trilinos project. ACM Trans Math Softw 31(3):397–423

24. Hindmarsh AC, Brown PN, Grant KE, Lee SL, Serban R, Shumaker DE, Woodward CS (2005) Sundials: suite of nonlinear and differential/algebraic equation solvers. ACM Trans Math Softw 31(3):363–396

25. Jiang Z, Li J, Yang Y, Mu L, Wei C, Yu X, Pianetta P, Zhao K, Cloetens P, Lin F, et al (2020) Machine-learning-revealed statistics of the particle-carbon/binder detachment in lithium-ion battery cathodes. Nat Commun 11(1):1–9

26. Li W, Sengupta N, Dechent P, Howey D, Annaswamy A, Sauer DU (2021) Online capacity estimation of lithium-ion batteries with deep long short-term memory networks. J Power Sources 482:228863

27. Li W, Zhang J, Ringbeck F, Jöst D, Zhang L, Wei Z, Sauer DU (2021) Physics-informed neural networks for electrode-level state estimation in lithium-ion batteries. J Power Sources, 506:230034

28. Long JW, Dunn B, Rolison DR, White HS (2004) Three-dimensional battery architectures. Chem Rev 104(10):4463–4492

29. Magnussen OM, Groß A (2019) Toward an atomic-scale understanding of electrochemical interface structure and dynamics. J Am Chem Soc 141(12):4777–4790

30. Naguib M, Allu S, Simunovic S, Li J, Wang H, Dudney NJ (2018) Limiting internal short-circuit damage by electrode partition for impact-tolerant li-ion batteries. Joule 2(1):155–167

31. Pannala S (2010) Computational gas-solids flows and reacting systems: theory, methods and practice: theory, methods and practice. IGI Global, Pennsylvania

32. Pannala S, Turner JA, Allu S, Elwasif WR, Kalnaus S, Simunovic S, Kumar A, Billings JJ, Wang H, Nanda J (2015) Multiscale modeling and characterization for performance and safety of lithium-ion batteries. J Appl Phys 118(7):072017

33. Pietsch P, Ebner M, Marone F, Stampanoni M, Wood V (2018) Determining the uncertainty in microstructural parameters extracted from tomographic data. Sustain Energy Fuels 2(3):598–605

34. Qian G, Zhang J, Chu S-Q, Li J, Zhang K, Yuan Q, Ma Z-F, Pianetta P, Li L, Jung K, et al (2021) Understanding the mesoscale degradation in nickel-rich cathode materials through machine-learning-revealed strain–redox decoupling. ACS Energy Lett 6(2):687–693

35. Qu X, Jain A, Rajput NN, Cheng L, Zhang Y, Ong SP, Brafman M, Maginn E, Curtiss LA, Persson KA (2015) The electrolyte genome project: A big data approach in battery materials discovery. Comput Mat Sci 103:56–67

36. Roberts M, Johns P, Owen J, Brandell D, Edstrom K, El Enany G, Guery C, Golodnitsky D, Lacey M, Lecoeur C, et al (2011) 3d lithium ion batteries—from fundamentals to fabrication. J Mat Chem 21(27):9876–9890

37. Roman D, Saxena S, Robu V, Pecht M, Flynn D (2021) Machine learning pipeline for battery state-of-health estimation. Nat Mach Intel 3(5):447–456

38. Saad Y, Schultz MH (1986) GMRES: a generalized minimal residual algorithm for solving nonsymmetric linear systems. SIAM J Sci Stat Comput 7(3):856–869

39. Scharf J, Chouchane M, Finegan DP, Lu B, Redquest C, Kim M-c, Yao W, Franco AA, Gostovic D, Liu Z, et al (2021) Bridging nano and micro-scale x-ray tomography for battery research by leveraging artificial intelligence. Preprint. arXiv:2107.07459

40. Severson KA, Attia PM, Jin N, Perkins N, Jiang B, Yang Z, Chen MH, Aykol M, Herring PK, Fraggedakis D, et al (2019) Data-driven prediction of battery cycle life before capacity degradation. Nat Energy 4(5):383–391

41. Slattery JC (1972) Momentum, energy, and mass transfer in continua. McGraw-Hill, New York

42. St. John PC, Guan Y, Kim Y, Kim S, Paton RS (2020) Prediction of organic homolytic bond dissociation enthalpies at near chemical accuracy with sub-second computational cost. Nat Commun 11:2328

43. Sundararaman R, Schwarz K (2017) Evaluating continuum solvation models for the electrode-electrolyte interface: challenges and strategies for improvement. J Chem Phys 146(8):084111

44. Tu H, Moura S, Fang H (2021) Integrating electrochemical modeling with machine learning for lithium-ion batteries. Preprint arXiv:2103.11580

45. Wang CY, Gu WB, Liaw BY (1998) Micro-macroscopic coupled modeling of batteries and fuel cells I. Model development. J Electrochem Soc 145(10):3407–3417

46. Wang H, Leonard DN, Meyer III HM, Watkins TR, Kalnaus S, Simunovic S, Allu S, Turner JA (2020) Microscopic analysis of copper current collectors and mechanisms of fragmentation under compressive forces. Mat Today Energy 17:100479

47. Ward L, Dandu N, Blaiszik B, Narayanan B, Assary RS, Redfern PC, Foster I, Curtiss LA (2021) Graph-based approaches for predicting solvation energy in multiple solvents: open datasets and machine learning models. J Phys Chem A 125(27):5990–5998
48. Wen M, Blau SM, Spotte-Smith EWC, Dwaraknath S, Persson KA (2021) BonDNet: a graph neural network for the prediction of bond dissociation energies for charged molecules. Chem Sci 12:1858–1868
49. Yao K, Herr JE, Toth DW, Mckintyre R, Parkhill J (2018) The tensormol-0.1 model chemistry: a neural network augmented with long-range physics. Chem Sci 9:2261–2269
50. Yun K-S, Pai SJ, Yeo BC, Lee K-R, Kim S-J, Han SS (2017) Simulation protocol for prediction of a solid-electrolyte interphase on the silicon-based anodes of a lithium-ion battery: ReaxFF reactive force field. J Phys Chem Lett 8(13):2812–2818
51. Zheng H, Li J, Song X, Liu G, Battaglia VS (2012) A comprehensive understanding of electrode thickness effects on the electrochemical performances of li-ion battery cathodes. Electro Acta 71:258–265
52. Zuo Y, Chen C, Li X, Deng Z, Chen Y, Behler J, Csányi G, Shapeev AV, Thompson AP, Wood MA, Ong SP (2020) Performance and cost assessment of machine learning interatomic potentials. J Phys Chem A 124(4):731–745

Index

This is a U.S. government work and not under copyright protection in the U.S.;
foreign copyright protection may apply 2023
S. Santhanagopalan (ed.), *Computer Aided Engineering of Batteries*,
Modern Aspects of Electrochemistry 62,
https://doi.org/10.1007/978-3-031-17607-4

Ingram Content Group UK Ltd.
Milton Keynes UK
UKHW022217210323
418907UK00001B/22